Applied Science Review™

Physiology

Leonard V. Crowley, MD
Pathologist
Riverside Medical Center
Minneapolis;
Visiting Professor
College of St. Catherine, St. Mary's Campus
Minneapolis;
Adjunct Professor
Lakewood Community College
White Bear Lake, Minn.;
Clinical Assistant Professor of Laboratory
Medicine and Pathology
University of Minnesota Medical School
Minneapolis

Charol Abrams MS, MT(ASCP)SH, CLS(NCA)CLSpH
Lecturer
Thomas Jefferson University
Philadelphia;
CLS Consultant and Medical Writer/Editor

Springhouse Corporation
Springhouse, Pennsylvania

Staff

EXECUTIVE DIRECTOR, EDITORIAL

Stanley Loeb

DIRECTOR OF TRADE AND TEXTBOOKS

Minnie B. Rose, RN, BSN, MEd

ART DIRECTOR

John Hubbard

CLINICAL CONSULTANTS

Maryann Foley, RN, BSN

EDITORS

Nancy Priff, Diane Labus, Keith de Pinho

COPY EDITOR

Pamela Wingrod

DESIGNERS

Stephanie Peters (associate art director), Matie Patterson (senior designer)

COVER ILLUSTRATION

Scott Thorn Barrows

ILLUSTRATORS

Jacalyn Facciolo, Jean Gardner, Robert Neumann

MANUFACTURING

Deborah Meiris (manager), Anna Brindisi, T.A. Landis

Printed in the United States of America.
ASR3-040792

Library of Congress Cataloging-in-Publication Data
Crowley, Leonard V., 1926-
Physiology / Leonard V. Crowley, Charol Abrams.
p. cm. — (Applied science review)
Includes bibliographical references and index.
1. Human physiology—Outlines, syllabi, etc.
I. Title.
II. Series.
QP41.C86 1993
612—dc20
ISBN 0-87434-456-5 92-19911
CIP

Contents

Advisory Board

Reviewers

Acknowledgments and Dedications

I am happy to acknowledge the help and cooperation of Maryann Foley and the staff of Springhouse Corporation, who encouraged me to undertake this project, and my coauthor, Charol Abrams, who worked with me to bring this project to completion. I also would like to thank the many students in nursing and other health fields with whom I have been associated. Their comments and suggestions have helped me design a study guide to fill their needs.

To my wife, who was helpful and very understanding.
L.V.C.

I would like to thank the staff of Springhouse Corporation for their help and cooperation, especially Maryann Foley, who encouraged me to work with Dr.Crowley on this project and who provided support throughout its duration. I also would like to acknowledge all the allied health students with whom I have worked as an educator. Their need to clearly understand and use complex concepts encourage us to always find new ways to facilitate learning.

To all the students who use this book. And to my husband, Dan, and daughters Alicia, Emily, and Cara — for their patience and encouragement.
C.A.

Preface

This book is one in a series designed to help students learn and study scientific concepts and essential information covered in core science subjects. Each book offers a comprehensive overview of a scientific subject as taught at the college or university level and features numerous illustrations and charts to enhance learning and studying. Each chapter includes a list of objectives, a detailed outline covering a course topic, and assorted study activities. A glossary appears at the end of each book; terms that appear in the glossary are highlighted throughout the book in boldface italic type.

Physiology provides conceptual and factual information on the various topics covered in most physiology courses and textbooks and focuses on helping students to understand:

- the interrelationship between anatomy and physiology
- the structure and function of human cells
- nerve impulse transmission
- normal mechanisms and pathways in various body systems, including the nervous, musculoskeletal, sensory, respiratory, circulatory, lymphatic and immune, digestive, and endocrine systems
- the importance of water, electrolyte, and acid-base balance
- normal reproduction, including fertilization, pregnancy, labor, and lactation.

1

The Human Cell

Objectives

After studying this chapter, the reader should be able to:
- Define the cell and describe its main components.
- Discuss the functions of the main components and organelles of a typical cell.
- Explain how mitochondria convert food nutrients into the energy in adenosine triphosphate (ATP) and how ATP is broken down and recycled to provide cellular energy.
- Describe the ways in which substances can pass through a cell membrane.
- Explain how osmotic pressure changes can affect the cell.
- Explain how nuclear deoxyribonucleic acid (DNA) regulates cellular functions.
- Discuss the function of genes and the significance of heterozygous and homozygous gene pairs in dominant, recessive, codominant, and sex-linked genetic expression.
- Compare the types of ribonucleic acid (RNA) and their functions in protein synthesis.
- Identify and discuss the phases of cell division.
- Contrast mitosis and meiosis.

I. Cellular Components

A. General information

1. The cell is the structural and functional unit of all living matter; it is the smallest body structure that can perform all the fundamental activities of life, such as movement, ingestion, excretion, and reproduction
2. Although most cells are specialized, they all have these major components in common: *nucleus, cytoplasm,* and *organelles,* or little organs

B. Nucleus

1. The nucleus is the control center of the cell; it directs the activities of the cytoplasmic structures
2. The nucleus contains the genetic material of the cell
3. It also contains one or more *nucleoli,* spherical structures that synthesize RNA
4. A nuclear membrane separates the nucleus from the cytoplasm; pores in this membrane allow certain substances to pass through

C. Cytoplasm

1. The cytoplasm is a semifluid medium that surrounds the nucleus
2. It is surrounded by a semipermeable *plasma membrane (cell membrane)* that regulates the passage of certain materials in and out of the cell (see Section III, Substance Movement Across the Cell Membrane, for more information)
3. The cytoplasm contains various organelles and cellular secretory products

D. Organelles

1. Each organelle performs a specific function
2. Not all organelles are used to the same degree, depending on the function of a particular cell; for example, cells that synthesize large amounts of lipids have an abundant smooth endoplasmic reticulum, whereas cells that synthesize protein have a rough endoplasmic reticulum
3. The major organelles are the mitochondria, ribosomes, endoplasmic reticulum, Golgi apparatus (or complex), lysosomes, and centrioles
   a. *Mitochondria* are the energy-producing cellular structures
      (1) They contain enzymes that oxidize food nutrients
      (2) This oxidation produces ATP, which provides the energy for many cellular activities
   b. *Ribosomes* are nucleoprotein particles that are attached to the endoplasmic reticulum; protein synthesis takes place in these organelles
   c. *Endoplasmic reticulum* is a system of interconnecting tubular channels
      (1) Rough (granular) endoplasmic reticulum is covered with ribosomes
      (2) Smooth endoplasmic reticulum contains enzymes to synthesize lipids
      (3) Both types of endoplasmic reticula have fluid-filled channels that appear to connect all parts of the cytoplasm
   d. *Golgi apparatuses* synthesize carbohydrate molecules
      (1) These molecules combine with protein produced by rough endoplasmic reticulum
      (2) This combination forms secretory products, such as lipoproteins
   e. *Lysosomes* are digestive bodies that break down damaged or foreign material in the cell
      (1) Each lysosome is surrounded by a membrane separating its digestive enzymes from the rest of the cytoplasm; its enzymes digest materials brought into the cell by phagocytosis
      (2) The membrane of the lysosome fuses with that of the vacuoles, or cystoplasmic spaces, that have phagocytized material, allowing the lysosomal enzymes to digest the engulfed material
         (a) Digestion occurs in the phagocytic vacuole
         (b) Digestive enzymes do not escape into the cell cytoplasm
   f. *Centrioles* are short cylinders adjacent to the nucleus; during cell division, they move to opposite poles of the cell and form the mitotic spindle

II. Cellular Energy Generation

A. General information

1. Cellular activities require energy
2. Mitochondria are the cellular power stations
   a. They contain enzymes that oxidize food nutrients

Adenosine Triphosphate Structure

The illustration below shows the chemical structure of adenosine triphosphate (ATP).

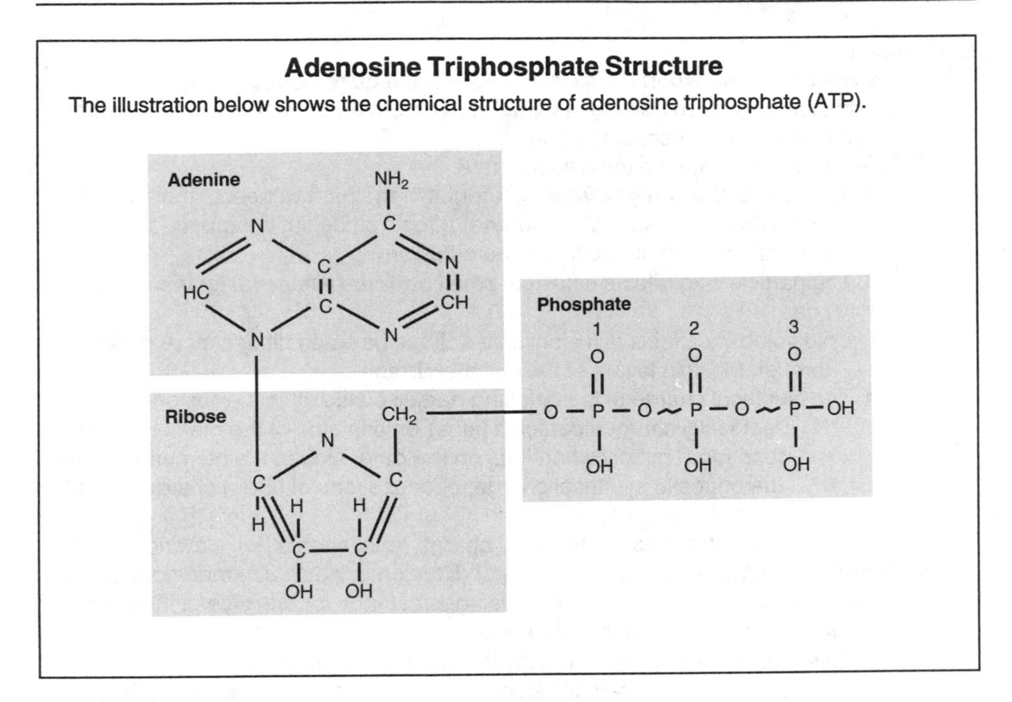

b. This oxidation produces ATP, a chemical fuel for cellular processes

B. Adenosine triphosphate (ATP)

1. ATP is composed of a nitrogen-containing compound (adenine) joined to a five-carbon sugar (ribose), forming adenosine; adenosine is joined to three phosphate groups (see *Adenosine Triphosphate Structure* for an illustration)
2. The chemical bonds between the first and second and the second and third phosphate groups contain a large amount of energy
   a. When the terminal high-energy phosphate bond ruptures, ATP is converted to adenosine diphosphate (ADP)
   b. Liberation of the third phosphate releases the energy stored in the chemical bond
3. Mitochondrial enzymes reconvert ADP and the liberated phosphate to ATP
   a. Mitochondria obtain the energy needed for this reattachment by oxidizing food nutrients
   b. The recycled ATP then is available again for energy production

III. Substance Movement Across the Cell Membrane

A. General information

1. Each cell interacts with body fluids through the interchange of substances
2. Movement of substances between cells and body fluids is accomplished by several methods of transport: ***diffusion, osmosis, active transport,*** and *endocytosis*
3. Transfer of fluids and dissolved substances across capillaries into interstitial fluid is facilitated by filtration

B. Diffusion

1. This method of transport is *passive;* it does not require cellular energy
2. In diffusion, dissolved particles (solute) move from an area of higher concentration to one of lower concentration
3. Several factors influence the diffusion rate
   a. The ***concentration gradient*** (difference in the particle concentration on either side of the plasma membrane) affects diffusion: the greater the concentration gradient, the faster the diffusion
   b. The particle size affects diffusion; small particles diffuse faster than large ones
   c. Lipid solubility affects diffusion; lipid-soluble particles diffuse more rapidly through the lipid layers of the cell membrane
   d. The electrical charge of the diffusing particles also affects diffusion
      (1) Electrically charged particles (ions) on one side of the membrane diffuse more rapidly when ions on the other side of the membrane have the opposite electrical charge, because ions of unlike charges are attracted to each other
      (2) Ions with the same electrical charge repel each other, slowing diffusion
4. ***Facilitated diffusion*** is a special type of diffusion in which a carrier molecule in the cell membrane picks up the diffusing substance on one side of the membrane and deposits it on the other side
   a. This process occurs more rapidly than simple diffusion
   b. An example of facilitated diffusion is the transport of glucose into cells by a carrier molecule influenced by insulin

C. Osmosis

1. Like diffusion, osmosis is a ***passive transport*** method that involves molecule movement from a solution of higher molecular concentration to one of a lower concentration
2. Unlike diffusion, osmosis involves water (solvent) molecule movement across the cell membrane from a dilute solution (with a high concentration of water molecules) to a concentrated one (with a lower concentration of water molecules)
3. Osmosis depends on the ***osmotic pressure*** of a solution
   a. Osmotic pressure measures the "water-attracting" property of a solution; it is determined by the number of dissolved particles in a given volume of solution, not on their size or electrical charge
   b. For example, a calcium chloride molecule ($CaCl_2$) dissociates (ionizes) in solution into three particles, a calcium ion and two chloride ions; its osmotic pressure is higher than that of a larger glucose molecule, which does not dissociate into ions when dissolved in solution
4. Water movement in and out of cells by osmosis depends on the osmotic pressure differences between intracellular and extracellular fluids
   a. Normally, intracellular osmotic pressure equals the extracellular osmotic pressure; as a result, the water content of cells does not change
   b. Osmotic pressure changes in body fluids causes water to shift between cells and extracellular fluids, impairing or disrupting cell functions

(1) When the osmotic pressure of extracellular fluid is lower than that of intracellular fluid, water enters the cells, causing them to swell and possibly rupture

(2) Conversely, when the osmotic pressure of extracellular fluid is higher than that of intracellular fluid, water moves into extracellular fluid, causing the cells to shrink

D. Active transport

1. Active transport is a transport method in which a substance is moved across the cell membrane
   a. Usually, the transport mechanism moves a substance from an area of lower concentration to an area of higher concentration (against the concentration gradient)
   b. This mechanism also can move a substance from an area of higher concentration to one of lower concentration (with the concentration gradient)
2. In this process, a carrier molecule in the cell membrane combines with the substance, transports it through the membrane, and deposits it on the other side of the membrane
3. Unlike facilitated diffusion, which also uses a carrier molecule, active transport requires energy from ATP breakdown to transport a substance across a cell membrane

E. Endocytosis

1. In this *active* (energy-requiring) method of transport, a substance is engulfed by the cell rather than passing through the cell membrane
   a. The cell surrounds the substance with part of the cell membrane
   b. This part of the membrane separates, forming a vacuole that moves to the cell interior
2. Endocytosis may be divided into phagocytosis and pinocytosis
   a. ***Phagocytosis*** is the engulfment and ingestion of particles too large to pass through the cell membrane
      (1) The phagocytic vacuole fuses with a lysosome; lysosomal enzymes then digest the engulfed particles
      (2) In some cases, phagocytosis is highly selective; specific particles are engulfed after first combining with specific receptors on the cell membrane, through receptor-mediated phagocytosis
   b. *Pinocytosis* is similar to phagocytosis, except that it is used only to engulf substances in solution or very small particles in suspension

F. Filtration

1. Movement of fluid and dissolved substances across a cell membrane also can be accomplished by filtration
2. In this transport method, the application of pressure to a solution on one side of the cell membrane forces fluid and dissolved particles through the membrane; the filtration rate depends on the amount of pressure
3. Filtration serves several purposes

a. It promotes the transfer of fluids and dissolved materials from the blood across the capillaries into the ***interstitial fluid*** (fluid surrounding the cells); the pressure of capillary blood provides the filtration force
b. Urine formation depends on fluid filtration from the blood flowing through capillaries in the kidneys (see Chapter 10, Urinary System, for more information)

IV. Cellular Genetic Material

A. General information

1. Chromosomes in the nucleus control cell activities
2. Chromosomes contain the genetic material of the cell

B. Chromosomes

1. *Chromosomes* control cellular activities; they direct protein synthesis by ribosomes in the cytoplasm
   a. Chromosomes are composed of DNA and protein
   b. They appear as a network of granules (chromatin) in the nondividing cell
2. Chromosomes exist in pairs, except in the *gametes* (male and female reproductive cells); one chromosome from each pair comes from the male parent, the other from the female parent
3. Normal cells contain 23 pairs of chromosomes
   a. In these cells, 22 pairs are sets of homologous chromosomes, or autosomes; they contain genetic information that controls the same characteristics or functions
   b. One pair is composed of sex chromosomes
      (1) The composition of these chromosomes (X and Y) determines sex
         (a) XX is genetic female
         (b) XY is genetic male
      (2) In the female, the genetic activity of both X chromosomes is essential only during the first few weeks of embryonic development
         (a) Later development requires only one functional X chromosome
         (b) The other X chromosome is inactivated and appears as a dense chromatin mass called a *Barr body* (or sex chromatin body), which is attached to the nuclear membrane in the cells of a normal female
      (3) The Barr body is absent in the cells of a normal male (who has only one functional X chromosome)

C. Genes

1. *Genes* are segments of chromosomal DNA chains that determine a cellular property; they are arranged in a line on the chromosomes somewhat like beads on a string
2. The *gene locus* refers to the location of a specific gene on a chromosome
   a. *Alleles* are alternate forms of a gene that can occupy a particular locus on a chromosome

b. Only one allele can occupy a specific gene locus

3. Because chromosomes are paired, genes also occur in pairs on homologous chromosomes, with one allele at its locus on both homologous chromosomes
   a. If the alleles for a particular gene are the same on both chromosomes, the individual is ***homozygous*** for that gene
   b. If the alleles are different, the individual is ***heterozygous*** for the gene
4. Genes are responsible for inherited traits; the effects they produce vary with the gene
5. *Gene expression* refers to the effect of a gene on cell structure or function
   a. A *dominant gene* is expressed in the heterozygous state
      (1) A dominant gene is expressed even if only one parent transmits it to the offspring
      (2) The genes for dark hair and eyes are dominant
   b. A *recessive gene* is expressed only in the homozygous state
      (1) A recessive gene is expressed only when both parents transmit it to the offspring
      (2) The genes for blond hair and blue eyes are recessive
   c. *Codominant genes* allow expression of both alleles, as in the genes that direct specific types of hemoglobin synthesis in red blood cells
   d. *Sex-linked genes* are carried on sex chromosomes; almost all appear on the X chromosome and are recessive in the male; sex-linked genes behave like dominant genes because there is no second X chromosome

D. Deoxyribonucleic acid (DNA)

1. *DNA*, a large nucleic acid molecule found in the chromosomes in the nucleus, is the carrier of genetic information in living cells
2. The basic structural unit of DNA is the *nucleotide;* it consists of a phosphate group linked to a five-carbon sugar, deoxyribose, joined to a nitrogen-containing compound called a base
   a. Deoxyribose is similar to ribose, except for the substitution of a hydrogen (H) atom for hydroxyl (OH) connected to a carbon (C) atom
   b. Four different DNA bases exist
      (1) Adenine and guanine are double-ring compounds classified as purines
      (2) Thymine and cytosine are single-ring compounds classified as pyrimidines
   c. Consequently, four different DNA nucleotides exist, differing only in the base joined to deoxyribose
3. Nucleotides are joined into long chains by chemical bonds between the phosphate group of the nucleotide and a carbon atom in the deoxyribose molecule of the adjacent nucleotide
4. The nitrogen bases project from the deoxyribose molecule at right angles to the long axis of the chain
5. DNA chains exist in pairs; they are held together by weak chemical attractions (hydrogen bonds) between the nitrogen bases on adjacent chains
   a. Because of the chemical configuration of the bases, adenine bonds only with thymine, and guanine bonds only with cytosine; bases that can link with each other are *complementary*

b. Linked DNA chains form a spiral structure, or double helix, which resembles a spiral staircase
   (1) The deoxyribose and phosphate groups form the railings
   (2) The nitrogen base pairs form the steps

6. The sequence of nucleotide bases in DNA chains forms the *genetic code,* a series of coded messages
7. Each group of three bases, called a *codon,* specifies the synthesis of a specific amino acid by attracting a specific amino acid, which is carried to the ribosomes to synthesize protein

V. Ribonucleic Acid (RNA)

A. General information

1. RNA transfers genetic information from nuclear DNA to ribosomes in the cytoplasm where protein synthesis occurs; several types of RNA are involved in this process
2. Like DNA, RNA consists of nucleotide chains; however, some of its components are different
   a. RNA contains the five-carbon sugar ribose rather than deoxyribose
   b. Uracil —not thymine —is the complemetary base of adenine in RNA
   c. Guanine and cytosine are complementary bases in RNA

B. Types of RNA

1. The nucleus produces three types of RNA, which pass into the cytoplasm: ribosomal RNA (rRNA), messenger RNA (mRNA), and transfer RNA (tRNA)
2. *rRNA* is used to make ribosomes in the endoplasmic reticulum of the cytoplasm, where the cell produces proteins
3. *mRNA* specifies how the amino acids are arranged to make proteins at the ribosomes
   a. mRNA consists of a single strand of nucleotides that is complementary to a segment of the DNA chain that contains instructions for protein synthesis
   b. The mRNA chains pass from the nucleus into the cytoplasm and attach to ribosomes in the cytoplasm
4. *tRNA* consists of short nucleotide chains; each chain is specific for an individual amino acid

C. Protein synthesis in the cytoplasm

1. One or more ribosomes attach themselves to the mRNA strand that contains instructions for protein synthesis
2. tRNA chains attach themselves to amino acids in the cytoplasm and transfer them to the ribosomes
3. The ribosomes join the amino acids into a chain according to the sequence of bases on the mRNA strand to form a protein
4. When the chain is complete, the new protein is released and the mRNA strand detaches from the ribosomes

VI. Cell Division

A. General information

1. Each cell must replicate itself for life to continue
2. Before a cell divides, its chromosomes are duplicated
3. Duplication involves separation of the DNA chains
4. Cells divide by ***mitosis*** or ***meiosis***
5. Although cell division is continuous, it may be divided into four phases: prophase, metaphase, anaphase, and telophase

B. Chromosome and DNA duplication

1. During this process, the double helix separates into two DNA chains; each serves as a template for constructing a new chain
2. Individual DNA nucleotides are linked into new strands with bases complementary to those in the originals
3. In this manner, two identical double helices are formed, each containing one of the original strands and a newly formed complementary strand
4. These double helices are duplicates of the original DNA chain (see *DNA Duplication,* page 10, for an illustration)

C. Mitosis

1. All cells except gametes undergo this form of cell division
   a. The parent cell, containing 46 chromosomes, undergoes division and gives rise to two daughter cells
   b. Both daughter cells are identical to each other and to the parent cell
2. During *prophase,* the first active phase of mitosis, the chromosomes (each consisting of two chromatids) thicken and shorten; the centrioles move to opposite sides of the cell and form the mitotic spindle, a network of protein microfibers; and the nuclear membrane breaks down
3. During *metaphase,* the chromosomes line up in the center of the cell; the two chromatids of each chromosome begin to separate but remain joined at the centromere, where spindle fibers are attached
4. During *anaphase,* the chromatids of each chromosome separate and are pulled to opposite poles of the cell by spindle fibers
5. During *telophase,* nuclear membranes form around both groups of chromosomes, and the cytoplasm divides, forming two identical daughter cells

D. Meiosis

1. Only gametes (ova and spermatozoa) undergo this type of cell division
2. It is characterized by intermixing of genetic material between homologous chromosomes and a reduction by half in the number of chromosomes in the four daughter cells
3. Meiosis consists of two separate divisions
4. The first meiotic division involves four phases
   a. During prophase, homologous chromosomes align so that matching genes are side by side, a condition called *synapsis;* chromatid segments may break off and interchange, or ***cross over,*** mixing the genetic material
   b. During metaphase, the chromosomes move to the center of the cell and the spindle fibers form

DNA Duplication

The basic structural unit of deoxyribonucleic acid (DNA) is the nucleotide, which is composed of a phosphate group, deoxyribose, and a nitrogen base made of adenine (A), guanine (G), thymine (T), or cytosine (C). Many nucleotide strands become twisted to form a double helix of a DNA molecule. During duplication, linked DNA chains separate. Then new complementary chains form and link to the originals (parents). This results in two identical double helices, consisting of parent and daughter, as shown.

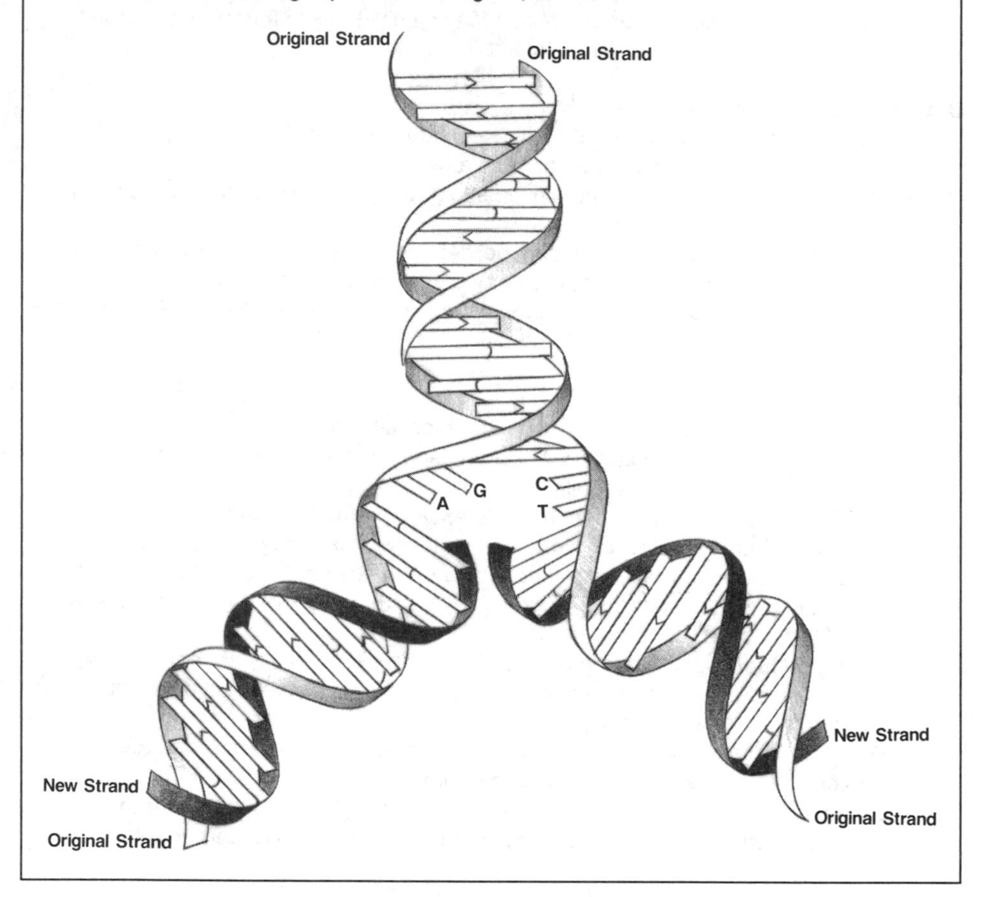

c. In anaphase, the homologous chromosomes —but not the chromatids —of each pair separate and move to opposite poles of the cell

d. During telophase, nuclear membranes form around the chromosomes, and the cytoplasm divides, forming two new daughter cells
   (1) Each cell has only 23 chromosomes
   (2) Each contains genetic material from both parents because of cross over and because the chromosomes of each parent do not all move to one side of the cell during anaphase

5. The second meiotic division resembles the four phases of mitotic division
   a. In each cell, the two chromatids of each chromosome separate to form new daughter cells

   b. However, because each cell that enters the second meitotic division has only 23 chromosomes, each daughter cell formed during this division has only 23 chromosomes
6. At the end of the second meiotic division, each parent cell has produced four daughter cells genetically different from the parent cell

Study Activities

1. Identify the major components of the cell and describe their functions.
2. Describe the process of cellular energy generation.
3. Compare and contrast the active and passive methods of transport across the cell membrane.
4. Discuss the three basic types of cellular genetic material.
5. Differentiate among the three types of RNA.
6. Trace the steps of mitosis and meiosis.

2

Nervous System

Objectives

After studying this chapter, the reader should be able to:

- Explain the general organization of the nervous system.
- Trace cerebrospinal fluid (CSF) movement through the central nervous system (CNS) and explain how constant CSF pressure is maintained.
- Compare and contrast the functions of the sympathetic and parasympathetic divisions of the autonomic nervous system.
- Describe the structure of nerve tissue.
- Compare and contrast nerve impulse transmission in myelinated and unmyelinated nerves.
- Explain how nerve impulses cross a synapse (contact point between neurons).
- Compare and contrast five neurotransmitters, noting their sources, functions, and modes of inactivation.
- State the major functions of the cerebrum, cerebellum, and brain stem.
- Explain how reflexes are generated and review the four different types.

I. Nervous System Divisions and Structures

A. General information

1. The nervous system directs every body system and governs all movement, sensation, thought, and emotion
2. It has two major components
   a. The *central nervous system* (CNS), which includes the brain and spinal cord, acts as the control center
   b. The *peripheral nervous system* (PNS), which includes the autonomic and somatic nervous systems, links the CNS with the rest of the body, and regulates voluntary and involuntary functions
3. Two types of nerve cells make up the nervous system, carrying electrical impulses throughout the body and stimulating the muscles and organs
   a. *Neurons* transmit impulses by receiving sensory stimuli, transmitting motor responses, or coordinating communication between body parts
   b. Four types of *neuroglia* (astroglia, ependymal cells, microglia, and oligodendroglia) support the neurons in special ways

B. The CNS

1. The CNS consists of the brain and spinal cord
   a. The *brain* is the center of the nervous system and controls all thoughts and actions
   b. The *spinal cord* serves as the primary pathway for messages traveling between peripheral areas of the body and the brain
2. The brain and spinal cord are surrounded by three membranes called *meninges*
   a. The *dura mater* is the fibrous outer membrane
   b. The *arachnoid membrane* is the middle membrane
   c. The *pia mater* is the inner membrane
3. The areas between the membranes form actual or potential spaces
   a. The *epidural space* lies over the dura mater
   b. The *subdural space* lies between the dura mater and the arachnoid membrane
   c. The *subarachnoid space* lies between the arachnoid membrane and pia mater and contains cerebrospinal fluid (CSF)
4. The carotid and vertebral arteries supply blood to the brain, which is vital to normal functioning; insufficient blood flow eventually leads to brain damage
   a. These arteries enter the brain through the base of the skull
   b. The vessels join to form a circle at the base of the brain (the *circle of Willis*)
   c. Smaller arteries extend from the circle of Willis to supply blood to all parts of the brain
5. Venous blood from the brain drains into large venous channels in the dura mater, which empty into the jugular vein
6. The interior of the brain contains four interconnecting channels called *cerebral ventricles* (two lateral ventricles, a third ventricle, and a fourth ventricle); these ventricles are filled with CSF secreted by the epithelial cells of the choroid plexuses (groups of blood vessels lining the ventricles)
7. CSF cushions the brain within the skull, protecting it from damage
   a. CSF flows through the ventricular system of the brain, exits through openings in the roof of the fourth ventricle, and circulates around the brain and spinal cord (see *Cerebrospinal Fluid Circulation,* page 14)
   b. CSF is absorbed into the venous sinuses through the *arachnoid villi* or *granulations* (tufts of arachnoid membrane that project into the large venous sinuses located in the dura mater)
   c. CSF also is absorbed directly into the veins on the surface of the brain
   d. Normally, the rates of CSF production and absorption are balanced so the CSF volume remains constant and the CSF pressure does not rise above normal

C. The PNS

1. The PNS consists of nerves that carry sensory (afferent) impulses to the CNS and motor (efferent) impulses to skeletal muscles, glands, and organs; its two

Cerebrospinal Fluid Circulation

Normally, cerebrospinal fluid (CSF) production is balanced with CSF absorption into the venous sinuses and into veins on the surface of the brain. In the illustration below, the arrows indicate the direction of CSF flow, which runs the full length of the spinal cord.

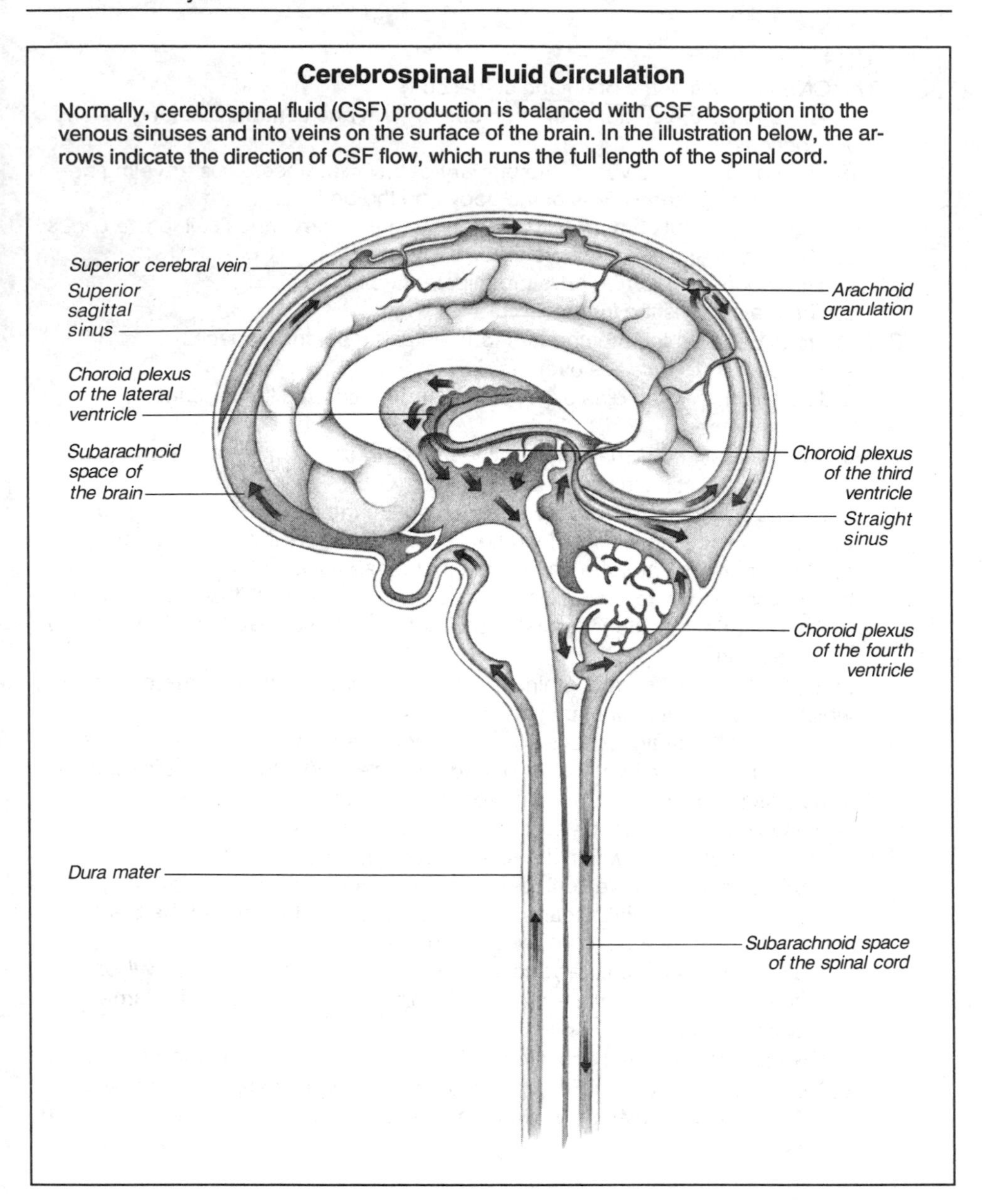

divisions the autonomic and somatic nervous systems control different functions

2. The *autonomic nervous system* (ANS) controls the involuntary functions of the smooth muscle, cardiac muscle, and glands (collectively called *effector organs*)

Responses to Autonomic Nervous System Stimulation

Most effector organs are innervated by the parasympathetic and sympathetic divisions of the autonomic nervous system. These divisions usually produce opposite responses as shown in the examples below.

EFFECTOR ORGANS	PARASYMPATHETIC RESPONSES	SYMPATHETIC RESPONSES
Eye		
Radial muscle of iris	None	Contraction (mydriasis)
Sphincter muscle of iris	Contraction for near vision	None
Heart	Decreased rate and contractility	Increased rate and contractility
Lung (bronchial muscle)	Contraction	Relaxation
Stomach		
Motility and tone	Increased	Decreased (usually)
Sphincters	Relaxation	Contraction (usually)
Intestine		
Motility and tone	Increased	Decreased
Sphincters	Relaxation	Contraction
Urinary bladder		
Bladder muscle	Contraction	Relaxation
Trigone and sphincter	Relaxation	Contraction
Skin		
Erector pilli	None	Contraction
Sweat glands	Generalized secretion	Slight localized secretion
Adrenal medulla	Secretion of epinephrine and norepinephrine	None
Liver	None	Glycogenolysis
Pancreas (acini)	Increased secretion	Decreased secretion
Adipose tissue	None	Lipolysis
Juxtaglomerular cells	None	Increased renin secretion

a. The ANS consists of two divisions, called the sympathetic and parasympathetic nervous systems; most ANS effector organs are supplied by both divisions, which usually have opposite effects on the organs (see *Responses to Autonomic Nervous System Stimulation* for examples of these effects)

b. The arrangement of neurons in the ANS differs from that in the somatic nervous system (see *Neuron Pathways in the Peripheral Nervous System* for a comparison)
   (1) In the ANS, impulses travel to the effector organs by means of a two-neuron chain, whereas the somatic system uses only one neuron
   (2) The cell body of the first neuron (preganglionic neuron) is in the CNS; its axon extends to a group of nerve cells called a *ganglion* outside the CNS
   (3) The cell body of the second neuron (postganglionic neuron) is in the ganglion outside the CNS; its axon extends from the ganglion to innervate the effector organ

c. The *sympathetic nervous system* usually assists the body in responding to stress

Neuron Pathways in the Peripheral Nervous System

The peripheral nervous system consists of the autonomic nervous system, which has parasympathetic and sympathetic divisions, and the somatic nervous system. Each of these systems has different neuron pathways, as shown below.

Autonomic nervous system pathways consist of two neurons: one neuron extends from the central nervous system (CNS) to a ganglion (preganglionic neuron); the other extends from the ganglion to the effector organ or gland (postganglionic neuron). Somatic nervous system pathways consist of only one neuron.

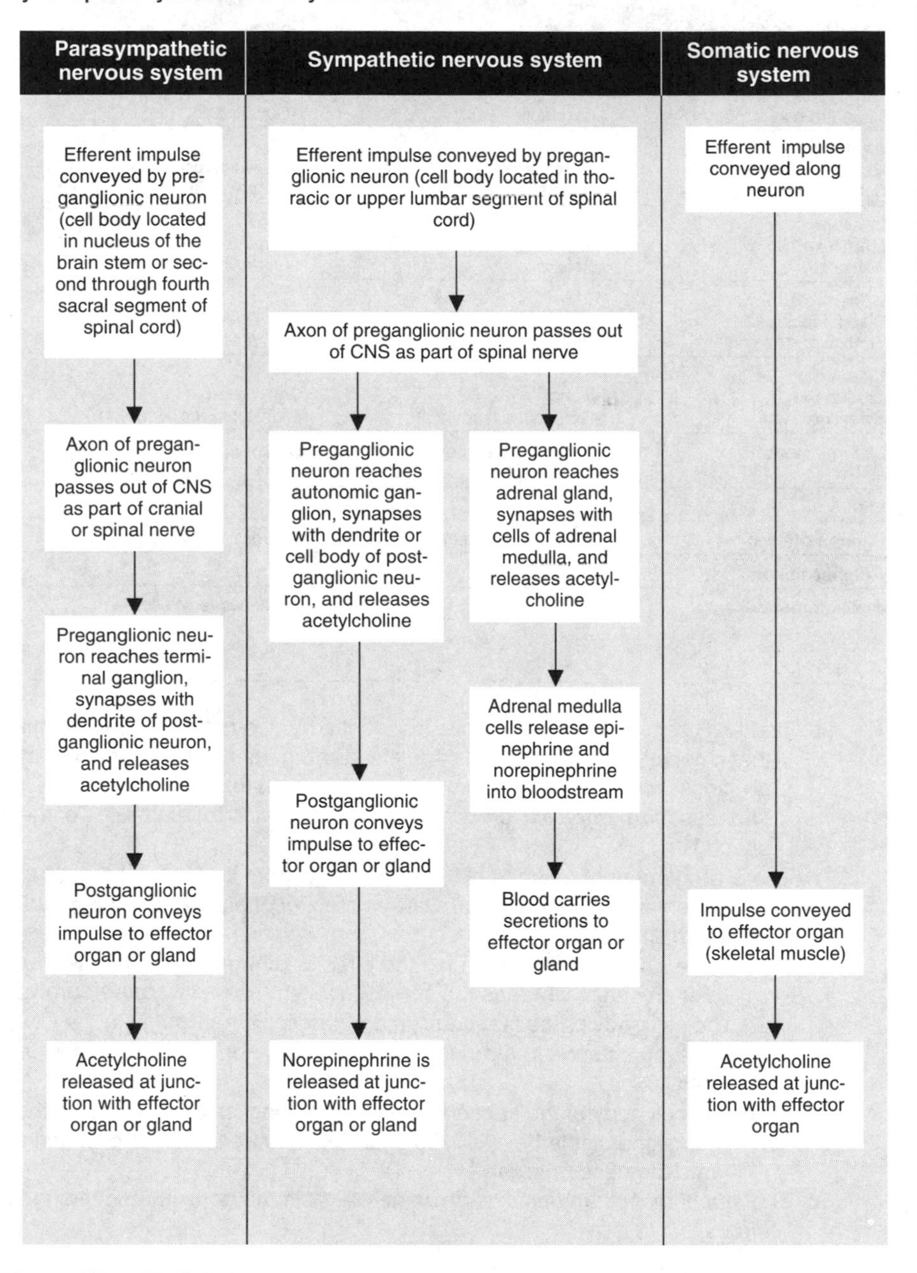

c. The *sympathetic nervous system* usually assists the body in responding to stress
   (1) Stimulation of the sympathetic nervous system prepares the individual to respond to stress (the *alarm response* or ***fight-or-flight response***)
   (2) In the sympathetic system, the ganglia are arranged in two chains in the thorax and abdomen adjacent to the intervertebral foramina (the sympathetic chains) and in clusters in front of the aorta and vertebral column (the prevertebral ganglia)

d. The *parasympathetic nervous system* typically produces the opposite effects of the sympathetic nervous system; it also deals with the day-to-day control of visceral effector muscles
   (1) Parasympathetic stimulation produces responses related to rest and relaxation
   (2) In the parasympathetic system, the ganglia are not grouped in distinct chains, but are located adjacent to or within the organs they supply

3. The *somatic nervous system* consists of cranial nerves and spinal nerves
   a. The 12 pairs of cranial nerves transmit sensory or motor impulses or both to the brain, controlling various functions (see *Cranial Nerve Functions,* page 18)
   b. The 31 pairs of spinal nerves transmit sensory and motor impulses to and from various body regions
      (1) Eight pairs of cervical nerves are called C1 through C8
      (2) Twelve pairs of thoracic nerves are called T1 through T12
      (3) Five pairs of lumbar nerves are called L1 through L5
      (4) Five pairs of sacral nerves are called S1 through S5
      (5) The final pair are called the coccygeal nerves
   c. The spinal nerves divide into anterior and posterior roots
      (1) Most of the anterior roots interconnect to form groups of nerves called *nerve plexuses,* which innervate the neck, arms, lumbar and sacral regions, and legs; however, the thoracic nerves innervate the chest and abdomen directly, without forming plexuses
      (2) The smaller posterior roots innervate the back muscles and overlying skin

D. Nerve tissue

1. Nerve tissue is composed of nerve cells called *neurons* and various types of supporting cells called *neuroglia,* such as astrocytes, ependymal cells, microglia, and oligodendroglia
2. Neurons are responsible for transmitting electrochemical impulses throughout the CNS and PNS; they do this by receiving sensory stimuli, transmitting motor responses, or integrating activities and coordinating communication between body parts
   a. *Sensory* (afferent) neurons transmit stimuli from peripheral sensory organs, such as the skin, to the spinal cord and brain
   b. *Motor* (efferent) neurons transmit impulses from the brain and spinal cord to peripheral organs and tissues
   c. *Interneurons* relay impulses within the CNS
3. Each neuron has a central body and two types of appendages; one or more dendrites and one long axon
   a. The *dendrites* receive impulses and transmit them toward the cell body

Cranial Nerve Functions

The twelve cranial nerves perform a wide range of sensory and motor functions, as described in the chart below.

NERVE NUMBER AND NAME	TYPE	FUNCTION
I (olfactory)	Sensory	• Provides a sense of smell
II (optic)	Sensory	• Provides vision
III (oculomotor)	Motor	• Moves the eyes and raises the eyelids • Adjusts the amount of light entering the eyes and focuses the lenses • Provides a sense of eye muscle movement
IV (trochlear)	Motor	• Moves the eyes • Provides eye muscle movement
V (trigeminal)	Motor and sensory	• Moves the muscles of mastication and those in the floor of the mouth • Provides sensory input from the eye surface, tear glands, upper eyelids, forehead, and scalp • Provides sensory input from the upper teeth, gingivae, and lip; palate lining; and facial skin • Provides sensory input from the lower teeth, gingivae, and lip; scalp; and skin over the jaw
VI (abducen)	Motor	• Moves the eyes • Provides eye muscle movement
VII (facial)	Motor and sensory	• Allows facial expression • Innervates tear and salivary glands • Provides a sense of taste to anterior tongue
VIII (vestibulocochlear)	Sensory	• Provides a sense of hearing • Provides a sense of equilibrium
IX (glossopharyngeal)	Motor and sensory	• Allows swallowing by controlling pharynx muscles • Innervates salivary glands • Provides a sense of taste to the posterior tongue • Provides sensory input from the pharynx, tonsils, and carotid arteries
X (vagus)	Motor and sensory	• Allows speech and swallowing by controlling pharynx muscles • Innervates muscles of the heart and smooth muscles of thoracic and abdominal organs • Provides sensory input from the pharynx, larynx, esophagus, and thoracic and abdominal organs
XI (spinal accessory)	Motor	• Allows movement of the soft palate, pharynx, and larynx • Allows movement of the neck and back
XII (hypoglossal)	Motor	• Allows tongue movement

b. The *axon* conducts impulses away from the cell body, through its presynaptic terminal, to adjoining neurons or effector organs

4. Most nerve fibers are surrounded by an insulating layer of phospholipid called *myelin*

a. Most nerve fibers that extend beyond the CNS are covered by elongated spindle cells called *Schwann cells,* which form a myelin sheath for the fibers; the sheath is interrupted by small gaps, called the *nodes of Ranvier*

b. Most nerve fibers within the CNS have a myelin sheath formed by extensions of oligodendroglial cells, which spiral around the fibers; the sheath of CNS fibers has fewer nodes of Ranvier than that of PNS fibers

c. Unmyelinated nerve fibers conduct impulses more slowly than myelinated ones

II. Neurotransmission

A. General information

1. *Neurotransmission* is the conduction of impulses throughout the nervous system; it occurs through the actions of neurons, which detect and transmit stimuli in the form of electrochemical impulses
2. Electrical transmission occurs within the nerve fiber
   a. Each neuron has an electrical potential ***(resting membrane potential)*** caused by different concentrations of sodium and potassium ions on each side of the membrane; normally the neuron is *polarized* (positive outside and negative inside)
   b. Stimulation of a nerve alters membrane permeability, allowing sodium to enter and suddenly causing the membrane to become *depolarized* (positive inside and negative outside)
   c. The spread of increased permeability and electrical current along the membrane is a nerve impulse
3. Chemical transmission occurs between two neurons or between a neuron and a muscle
   a. When the impulse reaches the end of a neuron, a *neurotransmitter* (a chemical transmitting substance) is released, which causes the impulse to cross the ***synapse***
   b. When the impulse is received by the next neuron, another chemical breaks down the neurotransmitter to prevent sustained impulse transmission
4. Sensory impulses ultimately reach the brain for interpretation; motor impulses are transmitted from the brain to the muscles or other effector organs

B. Transmission within nerve fibers

1. Polarization
   a. When a nerve fiber is not transmitting an impulse, it is in an electrical state called ***polarization***
      (1) The exterior of the nerve fiber has a positive charge; the interior has a negative charge
      (2) The voltage difference between the exterior and interior of the nerve fiber is approximately –70 millivolts (the minus sign indicates that the interior of the cells is negatively charged)
   b. This voltage difference, or *resting membrane potential,* is caused by the unequal distribution of sodium ions (Na^+) and potassium ions (K^+) on the two sides of the cell membrane
      (1) An active transport mechanism called the ***sodium-potassium pump*** transports potassium ions in through the membrane and simultaneously transports sodium ions out

(2) The pump carries in approximately two potassium ions for every three sodium ions carried out, causing a relative excess of positively charged ions on the exterior of the membrane

2. Depolarization
   a. Stimulation of the polarized nerve fiber causes ***depolarization*** (a change in resting membrane potential toward zero)
      (1) Stimulation causes a change in membrane permeability in the stimulated area
      (2) Positively charged sodium ions flow into the nerve fiber, enhancing depolarization
   b. Any stimulus strong enough to depolarize the membrane and initiate a nerve impulse will cause the depolarization wave, or impulse, to travel along the nerve fiber
   c. The membrane potential changes to zero and then reverses, with the interior of the nerve fiber having a positive charge relative to the exterior; this rapid change is called an *action potential*
      (1) The reversed polarity causes the membrane permeability to change again
         (a) Sodium ions are prevented from entering, while potassium ions are allowed to exit
         (b) This quickly returns the nerve fiber to its original polarity, with the interior being relatively negative in relation to the outside
      (2) Before the original polarization returns, however, the voltage difference between the area of reversed polarity and the adjacent polarized area of the membrane causes a current flow
         (a) This current flow depolarizes the adjacent area of the membrane; the action potential occurs there, causing a current flow to the next polarized area
         (b) The process continues along the entire length of the membrane, causing a depolarization wave that travels along the nerve fiber
         (c) As the nerve impulse travels in a depolarization wave, the previously depolarized area becomes repolarized and its resting membrane potential is restored
   d. In an unmyelinated nerve fiber, the impulse progresses without interruption along the entire length of the nerve fiber
   e. In a myelinated nerve fiber, impulse conduction is similar, but is influenced by the myelin sheath
      (1) The wave of depolarization "jumps" between the gaps in the myelin sheaths, bypassing the myelin-covered segments of the nerve fiber between the nodes of Ranvier
      (2) This transmission process, called ***saltatory*** *("jumping")* ***conduction,*** is much more rapid than the progressive depolarization wave that moves along an unmyelinated nerve fiber
   f. Nerve impulse transmission is followed by a brief interval, called the ***refractory period,*** when the nerve fiber is unresponsive and cannot transmit another impulse until its membrane is repolarized

C. Transmission across synapses

1. Nerve impulses are transmitted across synapses from neuron to neuron by chemicals that stimulate neurons
2. Neurons are organized in chains but are not in direct contact with each other
   a. Small gaps called *synapses* separate the neurons
   b. At the synapse, the axon terminal of the presynaptic neuron (the one transmitting the impulse) is close to the cell body or dendrite of the postsynaptic neuron (the next neuron in the chain) or to the muscle or organ it innervates
3. When the nerve impulse reaches the presynaptic axon terminal, it stimulates vesicles there to release a neurotransmitter
   a. The neurotransmitter diffuses across the synapse and binds to receptors on the membrane of the postsynaptic neuron; this changes the permeability of the postsynaptic neuron, which initiates membrane depolarization and impulse transmission
   b. The neurotransmitter released from the axon terminals is inactivated rapidly by various mechanisms to prevent a sustained response
4. Each type of neuron releases its own specific type of neurotransmitter
5. Because neurotransmitters are released only from axon terminals, transmission across a synapse is only possible from an axon to a dendrite, to the cell body of the next neuron, or to a muscle or other effector organ
6. Some neurons release *inhibitory neurotransmitters,* which inhibit, rather than stimulate, impulse transmission
   a. An inhibitory neurotransmitter causes the membrane of the postsynaptic neuron to become more permeable to potassium and chloride (Cl^-) ions without affecting the permeability to sodium
   b. Potassium ions diffuse out and chloride ions diffuse in, increasing the negative charge of the postsynaptic neuron membrane
   c. Then the membrane depolarizes less readily, which inhibits impulse transmission

III. Neurotransmitters and Neuropeptides

A. General information

1. Neurotransmitters are chemicals that are essential for neurotransmission; they are released by the axon terminal of a presynaptic neuron and travel across the synapse to excite or inhibit a target cell
2. After they are released, all neurotransmitters must be inactivated to prevent further stimulation
3. The chief neurotransmitters in the CNS are acetylcholine, norepinephrine, dihydroxyphenylethylamine (dopamine), 5-hydroxytryptamine (serotonin), and gamma-aminobutyric acid (GABA)
4. The chief neurotransmitters in the PNS are acetylcholine and norepinephrine
5. Neuropeptides regulate or modulate the actions of neurotransmitters

B. CNS neurotransmitters

1. Neurotransmitters are essential for neurotransmission in the CNS; other substances also may be involved

a. *Neuropeptides* are molecules composed of short chains of amino acids that usually appear in the axon terminals at synapses; they modify or regulate neurotransmitter activity
b. Other substances, such as the amino acids glycine and glutamic and aspartic acids, may play a role in neurotransmission, but their exact contribution is unknown

2. All neurotransmitters, neuropeptides, and related substances must be inactivated after they are released to limit their duration of activity
3. The chief neurotransmitters in the CNS are acetylcholine, norepinephrine, dihydroxyphenylethylamine (dopamine), 5-hydroxytryptamine (serotonin), and gamma-aminobutyric acid (GABA)
   a. Acetylcholine is stored in the synaptic vesicles of certain neurons, called *cholinergic neurons;* cholinergic neurons include all preganglionic neurons of the sympathetic and parasympathetic nervous systems, all postganglionic parasympathetic neurons, and postganglionic neurons that innervate sweat glands and some blood vessels
      (1) Acetylcholine is released in response to nerve stimulation and activates postganglionic neurons
      (2) It is inactivated rapidly by the enzyme cholinesterase, which breaks it down into acetate and choline
         (a) Some acetate diffuses away and is used by cells for energy
         (b) Axon terminals take up the choline, which then combines with acetate to form more acetylcholine
         (c) This acetylcholine is stored in synaptic vesicles for later release in response to further nerve stimulation
   b. Norepinephrine is stored in the synaptic vesicles of *adrenergic neurons,* which include most postganglionic neurons of the sympathetic nervous system
      (1) Norepinephrine is one of a group of neurotransmitters collectively called *catecholamines*
      (2) It is released from the axon terminals of most postganglionic neurons
      (3) Norepinephrine is inactivated in two ways
         (a) Some is taken back into the synaptic vesicles in the axon terminals *(reuptake)*
         (b) Some is inactivated by the enzymes catechol-O-methyltransferase (COMT) and monoamine oxidase (MAO)
   c. Dopamine is a precursor of norepinephrine and also is a catecholamine
      (1) This neurotransmitter stimulates postsynaptic neurons after being released from axon terminals
      (2) It is inactivated in the same two ways as norepinephrine
   d. Serotonin is a derivative of the amino acid tryptophane
      (1) It is stored in the synaptic vesicles of axon terminals
      (2) Serotonin is released in response to nerve stimulation and stimulates postsynaptic neurons
      (3) It is inactivated by reuptake and by enzymatic (MAO) breakdown
   e. GABA
      (1) GABA is an inhibitory neurotransmitter produced from glutamic acid, an amino acid
      (2) It is inactivated by enzymatic breakdown

4. Neuropeptides are closely related to neurotransmitters; they are chains of amino acids (peptides) that regulate or modulate the actions of neurotransmitters
   a. A neuropeptide called *substance P*, found in sensory nerves and in the spinal cord, facilitates the transmission of pain sensations to the brain
   b. Other neuropeptides, such as enkephalins, endorphins, and dynorphin, inhibit the perception of pain impulses transmitted to the brain

C. ANS neurotransmitters

1. Acetylcholine is a key neurotransmitter in the ANS, which is part of the PNS
   a. It is released from the axons of preganglionic neurons connected with the ANS, including all preganglionic neurons, postganglionic parasympathetic neurons, and postganglionic sympathetic neurons that innervate sweat glands and that cause vasodilation in certain blood vessels in skeletal muscle (sympathetic vasodilator nerves)
   b. It stimulates postganglionic neurons, which in turn release acetylcholine or the neurotransmitter norepinephrine from their axons
   c. As a postganglionic neurotransmitter, it also activates effector organs
   d. Acetylcholine is broken down rapidly by cholinesterase to prevent a sustained response (see "CNS neurotransmitters" for more information)
2. Norepinephrine also is a key neurotransmitter in the ANS; it is related structurally and functionally to epinephrine, which is produced by the adrenal medulla
   a. Most postganglionic sympathetic neurons release norepinephrine (except for the ones that release acetylcholine)
   b. After norepinephrine is released from nerve endings, some is inactivated by reuptake into axons and some by the action of COMT and MAO
   c. Because norepinephrine inactivation does not occur as rapidly as acetylcholine inactivation, the effects of sympathetic stimulation persist longer than those produced by parasympathetic stimulation
   d. Sympathetic stimulation also triggers release of epinephrine and norepinephrine from the adrenal medulla, which augments the effects of the norepinephrine produced by postganglionic sympathetic neurons
3. Neurotransmitter receptors exist in effector organs
   a. Acetylcholine released from postganglionic autonomic nerve endings produces its effect by combining with specific receptors on effector organs called *cholinergic receptors*
   b. When released from the postganglionic sympathetic neurons, norepinephrine produces its effects by combining with specific receptors on effector organs called *adrenergic receptors;* epinephrine and norepinephrine from the adrenal medulla produce similar effects on the same receptors
      (1) Adrenergic receptors are divided into two major groups: *alpha* and *beta receptors*
         (a) Alpha receptors contain two subgroups, $alpha_1$ and $alpha_2$
         (b) Beta receptors contain two subgroups, $beta_1$ and $beta_2$
      (2) Most effector organs contain alpha or beta adrenergic receptors, but some contain both
         (a) Organs that contain beta receptors commonly contain $beta_1$ and $beta_2$ receptors, but usually one type predominates
         (b) Blood vessel walls contain $alpha_1$ receptors, which cause vasoconstriction and tend to predominate over the counterbalancing ef-

fects of $alpha_2$ receptors, which are located primarily in terminals of postganglionic sympathetic neurons

(3) Usually, norepinephrine stimulates alpha adrenergic receptors

(4) Epinephrine stimulates $alpha_1$, $alpha_2$, $beta_1$, and $beta_2$ receptors

(5) Alpha and beta receptors produce different responses to adrenergic stimulation

(a) Stimulation of $alpha_1$ receptors produces contraction (vasoconstriction) of smooth muscle walls of blood vessels

(b) Stimulation of $alpha_2$ receptors produces the opposite effect by inhibiting norepinephrine release from sympathetic nerve endings

(c) Stimulation of $beta_1$ receptors, which predominate in cardiac muscle, cause the heart to beat faster and more forcefully

(d) Stimulation of $beta_2$ receptors, which predominate in the smooth muscle of bronchial walls and blood vessels, dilates bronchi and relaxes blood vessels

IV. Control Centers for Nervous System Function

A. General information

1. The control centers for most nervous system functions are located in the brain
2. Each of the three major areas of the brain the *cerebrum, cerebellum,* and *brain stem* controls specialized groups of functions
3. The spinal cord is primarily a reflex response center; it is modulated by the brain
4. Most simple ***reflexes*** result from neurotransmission through the ***reflex arc,*** a three-neuron chain composed of sensory, connecting, and motor neurons

B. The cerebrum

1. The cerebrum, which controls all advanced mental activities, is divided into two hemispheres
   a. The two cerebral hemispheres are divided into the frontal, parietal, temporal, and occipital lobes by large fissures
   b. The hemispheres are joined by a connecting bridge of nerve fibers, the *corpus callosum*
2. The *cerebral cortex,* which covers the surface of each hemisphere, is composed of gray matter that contains the cell bodies of neurons and white matter that contains their myelinated nerve fibers
3. Masses of gray matter called *basal ganglia* (the caudate and lentiform nuclei, claustrum, and amygdaloid nucleus) are located deep within each cerebral hemisphere; basal ganglia form part of the *extrapyramidal system,* which controls the coordination of muscle groups that function together to perform voluntary motion
4. The *internal capsule* is white matter, consisting of bundles of nerve fibers, that passes through the basal ganglia carrying sensory and motor impulses to and from the cerebral cortex
5. Certain parts of the cortex, called ***functional areas,*** are related to specialized functions; however, their functions are not sharply separated because of the extensive interconnections between the areas
6. The motor area controls voluntary motor activity
   a. The area contains neurons that control specific body parts

(1) Neurons in the upper part of the motor area control muscles in the lower part of the body
(2) Neurons in the lower part of the area control muscles in the head, neck, and upper part of the body
b. The number of neurons supplying a muscle depends on the type of movement it performs
(1) Muscles capable of fine movements, such as finger muscles, have large areas of cortical representation
(2) Muscles capable only of relatively gross movements, such as large limb and back muscles, are represented with smaller areas

7. The sensory area receives sensory impulses
   a. The upper part of the sensory cortex receives impulses from the lower part of the body
   b. The lower part of the sensory cortex receives impulses from the upper part of the body
8. Each cerebral hemisphere receives sensory impulses from or supplies motor impulses to the opposite side of the body because almost all the bundles of nerve fibers carrying impulses, or *fiber tracts,* cross to the opposite side of the brain, brain stem, or spinal cord as they ascend (or descend) the CNS
9. Cortical areas associated with vision are located in the occipital lobe; those associated with hearing are located in the temporal lobes adjacent to the lateral fissures
10. Motor areas related to speech are located in the frontal lobes; cortical areas related to olfactory sensations are located on the undersurface of the temporal lobes
11. Other cortical areas surrounding the primary areas are called *association areas;* although they are concerned with the same types of functions as the primary areas, they are more involved with interpretation, learning, and memory
12. The frontal areas primarily involve personality and judgment

C. The cerebellum

1. The cerebellum receives sensory impulses from muscles, joints, and tendons that convey a sense of position; it also receives impulses from the inner ear that involve balance and equilibrium
2. Cerebellar motor impulses regulate muscle groups that coordinate position and balance
3. The cerebellum is connected to the brain stem by bundles of fibers called the *cerebellar peduncles*
   a. Cerebellar peduncles transmit nerve impulses to the spinal cord, medulla, and brain
   b. The peduncles transmit impulses from the cerebellum to the thalamus for eventual transmission to the cortex

D. Brain stem

1. The brain stem is divided into the diencephalon, mesencephalon (midbrain), pons, and medulla; these structures control basic body functions
2. The diencephalon is the upper part of the brain stem and contains the slit-like third ventricle

a. The thalamus, which forms the lateral walls of the third ventricle, contains relay stations that receive sensory impulses and transmit them to the cortex

b. The hypothalamus contains neurons that control hormone output from the endocrine glands and regulate the activity of the autonomic nervous system; these neurons are responsible for such basic body functions as temperature regulation, food and water intake, and sexual behavior

3. The mesencephalon contains cell bodies of cranial nerves and large nerve fiber bundles that convey impulses to and from the cerebral hemispheres
4. The pons is a transverse bridge of fibers connecting the brain stem with the cerebellum; the pons also contains fiber tracts extending to and from the cerebral hemispheres, neurons of several of the cranial nerves, and neurons involved with spontaneous respiratory movements
5. The medulla is the lowest part of the brain stem; it forms the floor of the fourth ventricle, which is partially covered by the cerebellum
   a. The medulla contains nerve cell bodies of cranial nerves and nerve fiber bundles, which relay impulses to higher and lower levels in the nervous system
   b. It also contains neurons that regulate respiration, heart rate, and the degree of blood vessel contraction

E. Spinal cord

1. The spinal cord, which is the major reflex center in the CNS, mediates most reflexes
2. *Reflexes* are automatic actions, such as a knee jerk elicited by tapping the patellar tendon
3. Three common types of somatic (skeletal muscle) reflexes are the *stretch reflex* (involving two neurons and one synapse), the *flexor reflex* (involving sensory, motor, and association neurons and more than one synapse), and the *crossed-extensor reflex* (involving the flexor reflex and a contralateral reflex arc)
4. Most simple reflexes result from nerve impulse transmission through a three-neuron chain between sensory, connecting, and motor neurons; these neurons form the reflex arc (see *The Reflex Arc* for an illustration of a simple reflex)
   a. The sensory (afferent) neuron carries the impulse into the spinal cord through the dorsal root to the connecting neuron in the posterior horn of the spinal cord grey matter
   b. The connecting neuron relays the impulse to a motor neuron in the anterior horn of the grey matter
   c. The motor neuron sends out a motor (efferent) impulse via an axon in the ventral root of the spinal cord; this impulse causes muscle contraction
   d. Although a reflex arc is described in terms of a three-neuron chain, large numbers of sensory, connecting, and motor neurons are involved
      (1) The stimulus initiating the reflex usually activates a large number of sensory neurons, which transmit impulses to correspondingly large numbers of connecting neurons
      (2) These, in turn, cause discharge of many motor neurons to activate numerous muscle fibers
   e. In many reflexes, impulses initiated by sensory neurons also are relayed up and down the spinal cord by connecting neurons, causing a coordinated response from many different groups of motor neurons
5. Types of reflexes

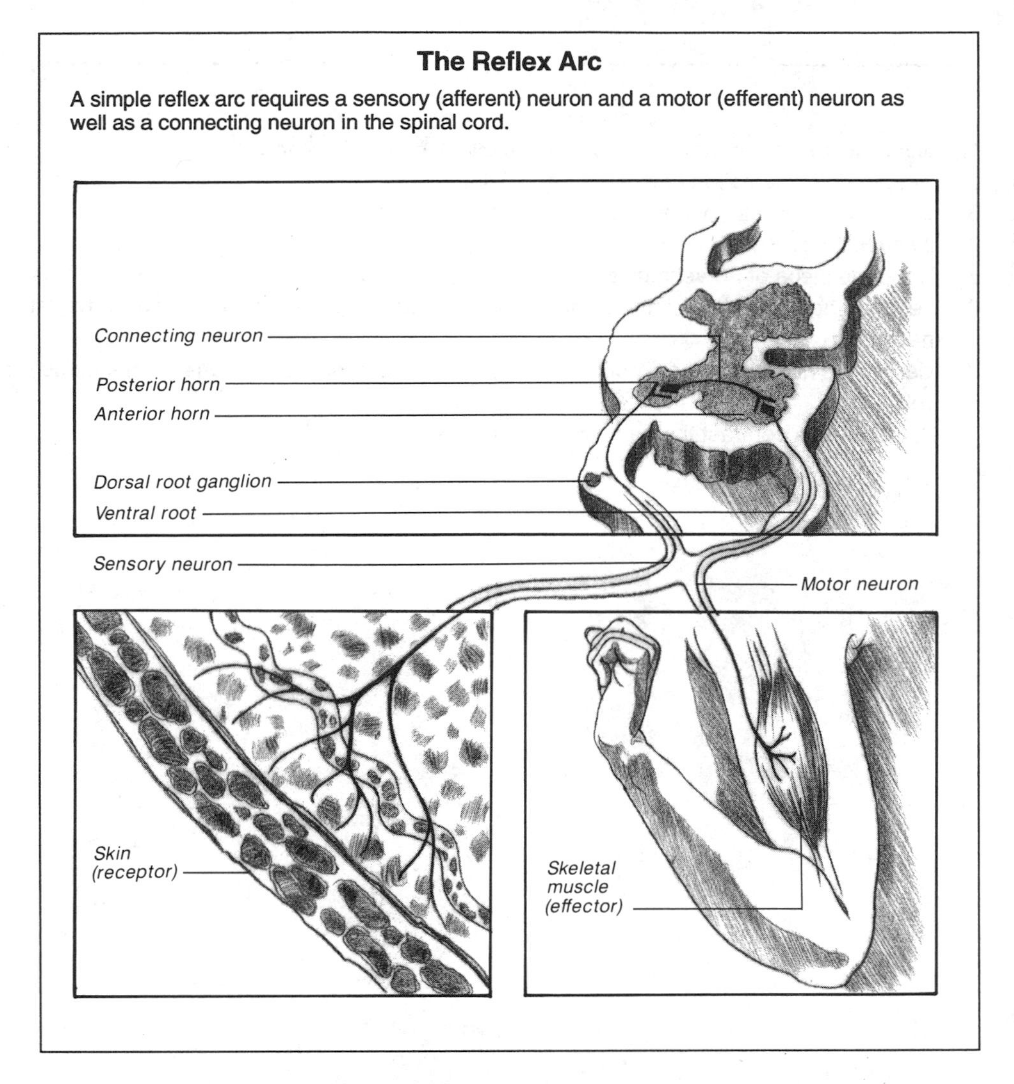

The Reflex Arc

A simple reflex arc requires a sensory (afferent) neuron and a motor (efferent) neuron as well as a connecting neuron in the spinal cord.

a. In many cases, the person is aware of the reflex response because the stimulus that initiated the reflex and the perception of the response are conveyed to the cortex; for example, the patellar reflex is automatic but also is perceptible

b. Some reflexes, such as those that regulate heart rate and blood pressure, function automatically; the person is not aware of the reflex control of these physiologic functions

c. Some reflexes can be inhibited or modified by conscious effort, such as control of urination and defecation reflexes

d. Some reflexes, called ***conditioned reflexes*** or *conditioned responses,* can be learned as a result of past associations, such as salivation induced by the sight or smell of specific foods

Study Activities

1. Differentiate among the structures and functions of the CNS and PNS.
2. Explain CSF circulation and pressure maintenance.
3. Identify the divisions of the PNS and describe their functions.
4. Discuss the functions of neurons and neuroglia in nerve tissue.
5. Trace the steps of nerve impulse transmission in myelinated and unmyelinated nerves.
6. Describe how neurotransmitters conduct an impulse across synapses and how they are inactivated.
7. Identify the nervous system control centers and explain which functions each area controls.
8. Compare and contrast the four types of reflexes.

3

Musculoskeletal System

Objectives

After studying this chapter, the reader should be able to:
- Identify the three major types of muscle and describe their functions.
- Describe nerve impulse transmission across the neuromuscular junction.
- Explain the arrangement of actin and myosin filaments and their function in skeletal muscle contraction.
- Define a motor unit and provide three examples.
- Compare the structural and functional characteristics of each type of muscle cell.
- Describe the energy sources for muscle contraction and explain how muscular activity can cause oxygen debt.
- Describe the two types of bone formation and the factors that affect them.
- Contrast bone formation and resorption.
- Compare the basic structure and function of the three types of joints.
- Identify the major functions of bones.

I. Muscle Tissue

A. General information

1. Muscles are responsible for voluntary movements and many involuntary activities; different types of muscle cells control different activities
2. Nerves regulate muscle function
3. Muscle is composed of elongated cells called *muscle fibers* that are specialized for contraction
4. The cytoplasm of muscle fibers is called the *sarcoplasm*
   a. The sarcoplasm contains *myofibrils,* bundles of filaments that are the contractile elements of the cell
   b. It is surrounded by a plasma membrane called the *sarcolemma*

B. Types of muscle

1. Each type of muscle —skeletal, cardiac, and smooth muscle —differs in function and structure
2. *Skeletal muscle* moves the body
   a. It is regulated by the central and peripheral nervous systems and is attached to the skeleton
   b. Skeletal muscle is under voluntary control
3. *Cardiac muscle,* the specialized muscle of the heart, contracts rhythmically to pump blood

Comparing Muscle Types

Skeletal, cardiac, and smooth muscles differ in their location, microscopic appearance, striations, type of control, contraction speed, and function as described in the chart below.

FEATURES	SKELETAL MUSCLE	CARDIAC MUSCLE	SMOOTH MUSCLE
Location	Attached to the skeleton	Heart	Walls of hollow organs, blood vessels, skin, and eyes
Appearance	• Cylindrical fibers that may extend the entire length of a muscle • Multiple peripherally located nuclei	• Cylindrical branched fibers • Single centrally located nucleus • Intercalated disks that join the cells to each other	• Spindle-shaped fibers • Single centrally located nucleus
Striations	Present	Present	Absent
Control	Voluntary	Involuntary	Involuntary
Contraction speed	Fast	Moderate	Slow
Function	Body movement	Blood pumping	Food movement through the gastrointestinal tract, urinary bladder emptying, uterine and tubal muscle contraction, regulation of blood vessel diameter, pupil size regulation, hair movement, and nipple erection

a. It is regulated by the autonomic nervous system
b. Cardiac muscle is not under voluntary control

4. *Smooth muscle* exists in the walls of hollow organs and in the blood vessels, skin, and eyes
   a. Like cardiac muscle, it is regulated by the autonomic nervous system
   b. Smooth muscle is not under voluntary control (see *Comparing Muscle Types* for more information)

C. Muscle fiber stimulation

1. All three types of muscle are controlled by nerve impulses (see Chapter 2, Nervous System, for detailed information); although they are stimulated and respond in somewhat different ways, skeletal muscle will be used as an example
2. Nerve fibers called axons transmit impulses from the central nervous system to muscle fibers
3. The axon terminates near a small depression on the surface of the muscle fiber called the *motor end-plate;* a small gap called the synapse separates the axon terminal and motor end-plate
4. The term ***neuromuscular junction*** refers to the motor end-plate and the axon terminal associated with it

D. Nerve impulse transmission to muscle fibers

1. The axon terminal has vesicles that contain the neurotransmitter acetylcholine

a. When a nerve impulse reaches the axon terminal, the vesicles release acetylcholine, which diffuses across the synapse and attaches to receptors on the motor end-plate
b. Acetylcholine generates a depolarization wave, or impulse, which stimulates muscle fiber contraction

2. The enzyme cholinesterase is present in the muscle membrane of the motor end-plate
   a. This enzyme rapidly breaks down the released acetylcholine into acetate and choline
   b. Acetylcholine breakdown prevents the nerve impulse from continuing to stimulate the skeletal muscle fiber and inducing a sustained contraction
3. The axon terminal takes up the choline, which then combines with acetate to form more acetylcholine
4. Acetylcholine is stored in axon terminal vesicles until another impulse stimulates its release

II. Skeletal Muscle

A. General information

1. Skeletal muscle fibers are long, cylindrical cells
2. Each cell has several nuclei located at the cell periphery, directly beneath the sarcolemma
3. Abundant mitochondria are dispersed in the sarcoplasm; they generate the energy required for cellular function
4. A network of membranous channels runs parallel to the long axis of the muscle cell and surrounds each myofibril; this network is called the *sarcoplasmic reticulum*
5. Tubelike extensions of the sarcolemma called *transverse tubules (T tubules)* extend into the sarcoplasm at right angles to the long axis of the cell and sarcoplasmic reticulum
6. The muscle fibers are divided into segments called *sarcomeres*
   a. The muscle fibers are divided by bands of protein called *Z lines*
   b. Z lines extend across muscle fibers at right angles to their long axis
7. Bundles of muscle fibers make up skeletal muscles
   a. Each muscle fiber is covered by a sheet of fibrous connective tissue called the *endomysium*
   b. Bundles of muscle fibers, or *fasciculi* are surrounded by a sheath of connective tissue covered by a fascia called the *perimysium*
   c. Bundles of fasciculi are covered by a sheath of connective tissue called the *epimysium* to make up an entire muscle
8. Skeletal muscles are anchored to the skeleton by extensions of the connective tissue sheath surrounding the muscle fibers
   a. The extensions may attach directly to a bone
   b. They may blend into a strong fibrous connection called a *tendon,* which is anchored to the bone
   c. Either arrangement allows the muscles, when stimulated to contract, to move the bone

B. Myofibril structure

1. A myofibril is composed of thin filaments of the protein *actin* and thick filaments of the protein *myosin*
   a. The ends of the actin filaments attach to Z lines, which partition the myofibrils into sarcomeres
   b. Myosin filaments are not attached to Z lines, but are held in proper alignment by delicate threads that interconnect the midportions of these filaments
   c. When actin and myosin filaments overlap, each myosin filament is surrounded by six actin filaments
2. The pattern of overlapping actin and myosin filaments and Z lines makes the muscle fibers appear striated with dark and light longitudinal bands
   a. The dark bands (A bands) contain myosin filaments and the overlapping ends of actin filaments
   b. The light bands (I bands) contain only actin filaments
   c. A narrow band (H zone) in the center of the A band contains only myosin filaments
   d. A thin band (M line) in the center of the H zone contains the fine stabilizing threads that interconnect the myosin filaments
3. Each actin filament consists of a double chain of actin protein surrounded by threadlike structures composed of two other proteins, *tropomyosin* and *troponin;* in an unstimulated muscle fiber, tropomyosin physically blocks actin from binding to myosin
4. Each myosin filament consists of myosin protein strands in a long rodlike structure; myosin extensions or heads project laterally from the rod where actin filaments overlap the myosin filament

C. Skeletal muscle stimulation

1. Skeletal muscle is under voluntary control
   a. It contracts in response to impulses transmitted from the central nervous system by motor nerves
   b. These voluntary impulses cause the muscles to contract in a synchronized fashion and cause skeletal movement
2. When the axon of a motor neuron enters the muscle, it splits into branches or terminals that transmit impulses to muscle fibers; each motor neuron controls a variable number of muscle fibers
3. A group of muscle fibers supplied by a single neuron is called a *motor unit*
   a. Muscles capable of fine movements, such as eyelid or finger movement, may have only a few muscle fibers in each motor unit
   b. Muscles capable of gross movements, such as buttock and leg movement, may have several hundred muscle fibers in each motor unit
4. Muscle fiber contraction of a motor unit is governed by the ***all or none response***
   a. If the nerve impulse is intense enough to stimulate contraction, all the muscle fibers in the motor unit contract
   b. If the impulse is not intense enough to stimulate contraction of all the muscle fibers in the motor unit, none contract
5. Contraction strength depends on the number of motor units activated
   a. Weak contractions activate only a few motor units
   b. Stronger contractions activate more motor units

D. Skeletal muscle contraction and relaxation

1. Skeletal muscle contraction involves calcium ion transport and nerve impulse transmission
   a. When a nerve impulse stimulates the muscle fiber at the neuromuscular junction, the impulse is propagated along the sarcolemma and transmitted through the T tubules into the interior of the muscle fiber
   b. The impulse causes the sarcoplasmic reticulum to release calcium ions
   c. Calcium ions diffuse into the sarcoplasm and bind to troponin on the threadlike structures around the actin filaments, changing the position of these structures; as a result, tropomyosin no longer can prevent actin and myosin from binding
   d. Myosin heads bind to the actin filaments, forming structures called ***cross bridges*** that pull the actin filaments toward the center of the sarcomeres; this results in muscle fiber shortening, or contraction
2. Muscle relaxation follows contraction
   a. When the motor unit no longer is stimulated, calcium ions separate from troponin and are transported back into the sarcoplasmic reticulum
   b. The threadlike structures return to their original positions
   c. Tropomyosin unbinds the cross bridges, and actin and myosin filaments separate, causing the sarcomeres (and muscle fibers) to lengthen
3. Energy for skeletal muscle contraction may come from several sources
   a. One energy source is the breakdown of adenosine triphosphate (ATP) in muscle fiber, which occurs simultaneously with calcium ion release from the sarcoplasmic reticulum
   b. A second energy source is the breakdown of *phosphocreatine,* a compound found in muscle and composed of phosphate joined to creatine by a high-energy bond
      (1) When this bond is broken, energy is released
      (2) Some of the released energy is used to resynthesize ATP, which then becomes available as an energy source for continued muscle contraction
   c. Energy from ATP and phosphocreatine breakdown can supply immediate energy for short-term skeletal muscle contraction, but sustained exertion requires energy from other sources
      (1) Muscle glycogen is broken down into glucose, which is metabolized to yield energy
      (2) Glucose that is delivered to muscle fibers by the blood (blood sugar) also is metabolized to yield energy
      (3) Fatty acid breakdown also supplies energy, which is delivered to muscle fibers by the blood
   d. If skeletal muscle fibers have sufficient oxygen, glucose is converted to pyruvic acid and then oxidized by enzymes in the mitochondria to yield carbon dioxide and water to produce ATP; this process is called ***aerobic metabolism*** (see Chapter 9, Nutrition, Digestion, and Metabolism, for more information on metabolism)
   e. During vigorous exertion, the lungs and circulatory system may not be able to provide sufficient oxygen to muscles for aerobic metabolism
   f. In such a case, muscle cells must derive energy through ***anaerobic metabolism,*** which breaks down muscle glycogen and blood glucose to lactic acid rather than pyruvic acid; anaerobic metabolism provides less ATP than aer-

obic metabolism, but it yields enough to sustain muscular activity (see Chapter 9, Nutrition, Digestion, and Metabolism, for more information on metabolism)
(1) Some lactic acid accumulates in the muscles, but most diffuses into the blood and is transported to the liver
(2) In the liver, lactic acid is reconverted to glucose, which is transported as blood glucose back to the muscles to be reused for energy

g. With anaerobic metabolism, the muscles develop an oxygen debt that must be repaid through extra oxygen consumption after exertion has ceased; the person must breathe rapidly after exertion until the muscles have obtained sufficient oxygen to restore them to their resting state

h. When oxygen supplies increase after exertion, the muscle fibers are restored to their normal resting state
(1) Depleted muscle glycogen is resynthesized from glucose
(2) Phosphocreatine is reformed
(3) Accumulated lactic acid in the muscle is reconverted to pyruvic acid and oxidized by the mitochondrial enzymes to carbon dioxide and water and ATP (see Chapter 9, Nutrition, Digestion, and Metabolism, for more information on metabolism)

E. Muscle tone

1. Normal muscle is not relaxed completely; it maintains a slight, sustained contractility called *muscle tone*, which is a reflex contraction in response to stretching
   a. When a muscle is stretched, receptors in muscles, joints, and tendons respond by sending impulses to the nervous system
      (1) These afferent impulses are transmitted to the spinal cord and cause discharge of motor impulses that make the muscle fibers contract
      (2) The amount of stretching of the muscle fibers regulates the amount of reflex contraction
   b. Motor neurons then transmit impulses to the muscle fibers, causing them to contract and resist the stretching force; muscle tone is maintained by the contraction of muscles with opposing actions
      (1) Flexor muscles respond to the pull of extensor muscles by contracting slightly
      (2) Extensor muscles, in turn, respond to the pull of flexors, resulting in a slight state of tension in all muscle groups
2. Muscle tone maintains posture by maintaining continuous contraction of the head, neck, and back muscles
3. Muscle tone around joints maintains joint stability

III. Cardiac Muscle

A. General information

1. Cardiac muscle, which makes up the heart, contracts rhythmically to pump blood
2. Like skeletal muscle fibers, cardiac muscle fibers are elongated cells divided into sarcomeres; they have a striated pattern
3. Each fiber has a single nucleus located in the center of the cell

4. The sarcoplasmic reticulum of the cardiac muscle fiber is less prominent than in skeletal muscle, and the T tubules are associated less closely with the sarcoplasmic reticulum
5. Cardiac muscle cells interconnect to form branching networks

B. Cardiac muscle contraction and relaxations

1. Cardiac muscle cells have the same arrangement of actin and myosin and employ the same general mechanism of contraction and relaxation as skeletal muscle cells; however, some variations exist, especially in the manner of stimulation
2. Cardiac muscle contractions occur at a slower rate and last longer than skeletal muscle contractions because the T tubules and sarcoplasmic reticulum are farther apart in cardiac muscles, making nerve impulse transmission less efficient
3. The blood supply to —and mitochondria in — cardiac muscle fibers are more abundant than those in skeletal muscle fibers
   a. Cardiac muscle fibers obtain sufficient ATP for energy from aerobic metabolism of nutrients in the mitochondria
   b. These fibers do not rely on anaerobic metabolism during increased activity and do not develop an oxygen debt as skeletal muscle fibers do
4. Cardiac muscle fibers are not organized into motor units like skeletal muscle fibers
   a. Interconnections called *intercalated disks* between the branching cardiac muscle fibers permit action potentials to pass from cell to cell
   b. These muscle fibers function as an integrated unit
5. The autonomic nervous system regulates cardiac rate and rhythm, but cardiac muscle fibers can generate impulses and contract rhythmically even when deprived of nervous stimulation
6. During cardiac muscle contraction, electrocardiography can measure the electrical activity associated with depolarization and repolarization of cardiac muscle (see Chapter 6, Cardiovascular System, for detailed information about cardiac contraction and electrical conduction)

IV. Smooth Muscle

A. General information

1. Smooth muscle fibers are spindle-shaped cells with tapered ends and a single central nucleus
2. The sarcoplasm of these fibers contains fewer actin and myosin filaments than do skeletal muscle fibers; because smooth muscle filaments are not organized into sarcomeres, the fibers do not appear striated
3. The sarcoplasmic reticulum is not developed well in smooth muscle fibers; these fibers do not have a network of T tubules, but cytoplasmic vacuoles that connect with the surface of the fibers may serve the same function
4. The sarcoplasm contains a network of *intermediate filaments* attached to cytoplasmic structures called *dense bodies;* dense bodies are the counterparts of Z lines in striated muscle because actin filaments attach to them

B. Types of smooth muscle

1. *Visceral (single-unit) smooth muscle* forms a continuous interconnected network of muscle fibers that function as a unit
   a. This type of smooth muscle is found in the walls of blood vessels, where it regulates their diameter
   b. It also forms the walls of hollow internal organs, such as those of the gastrointestinal, reproductive, and urinary tracts
   c. Wavelike contractions of smooth muscle propel substances through the gastrointestinal and reproductive tracts; contraction of smooth muscle in the bladder expels urine
2. *Multiunit smooth muscle* consists of individual muscle fibers
   a. These muscle fibers do not contract as a unit because each fiber is controlled by a separate motor neuron
   b. Fibers of this type are present in several areas, such as the walls of bronchi, where contraction causes the bronchi and bronchioles to constrict; the skin, where contraction causes the hairs to stand on end; and the eye, where contraction of the iris and ciliary body regulate pupil size and lens focus

C. Smooth muscle contraction and relaxation

1. Contraction and relaxation of smooth muscle is basically the same as in skeletal muscle; however some variations exist
2. Actin and myosin filaments slide together during contraction, pulling on the network of intermediate filaments and dense bodies; this causes the smooth muscle fibers to shorten
3. Smooth muscle contracts and relaxes more slowly than skeletal muscle does
4. Smooth muscle contraction requires less energy than skeletal muscle contraction
5. Smooth muscle fibers receive motor impulses from the autonomic nervous system; they are not under voluntary control

V. Skeleton

A. General information

1. The skeleton contains 203 separate bones
2. Bones are composed of specialized connective tissue that consists of a collagenous matrix impregnated with calcium phosphate and small amounts of calcium carbonate and other minerals
3. Bone formation and breakdown ***(resorption)*** occur continuously, even after the bones stop increasing in size
   a. Formation exceeds resorption during childhood and adolescence, allowing the bones to grow
   b. In young and middle adulthood, bone formation and resorption are balanced closely; bone size and density do not change
   c. In old age, bone breakdown exceeds formation, which causes a gradual decrease in bone density

4. Bone strength and thickness are related to stresses on the bone
   a. Heavy physical activity and weight bearing promote heavier and stronger bones by stimulating osteoblast formation and bone matrix production and by inhibiting osteoclast activity
   b. Reduced activity leads to loss of bone density and strength
5. Bones are connected to skeletal muscles and to each other by joints and other connective tissue
6. Bones support and protect the body, allow movement and hematopoiesis, and act as a mineral reservoir

B. Bone formation

1. Two types of bone formation exist: *endochondral* and *intramembranous; osteoblasts* are the active bone-forming cells in both types of formation
   a. These cells produce an organic matrix for bones and also secrete the enzyme alkaline phosphatase, which liberates phosphate ions from compounds at the site of bone formation
   b. These phosphate ions combine with calcium ions to form calcium phosphate
   c. Calcium phosphate is deposited in the bone matrix, causing the bone to become rigid
   d. Osteoblasts become incorporated into the bone as it forms; then they are transformed into *osteocytes,* relatively inactive mature bone cells
2. In ***endochondral bone formation,*** a cartilage model forms first; then osteoblasts invade the cartilage and convert it into bone
   a. Areas of bone formation are called *centers of ossification*
   b. In a long bone, ossification begins first in the shaft, or diaphysis; then centers of ossification form in the ends of the bone, or epiphyses
   c. The actively growing zone of cartilage between the diaphysis and epiphyses is called the *epiphyseal line*
3. ***Intramembranous bone formation*** occurs in many of the flat bones of the skull and in the clavicle; osteoblasts form these bones directly without a preliminary cartilage mass
   a. Bone-forming cells differentiate from precursor cells in connective tissue and begin to produce bone directly within the connective tissue
   b. The initial centers of ossification extend peripherally and eventually convert all the connective tissue to bone
4. Bone increases in length and thickness as an individual grows
   a. In endochondral formation, bone grows in length by continuous production of cartilage at the distal end of the epiphyseal plate and conversion of cartilage into bone in the proximal part of the epiphyseal plate; this is followed by remodeling of the bone to maintain its configuration
      (1) Toward the end of adolescence, the epiphyseal lines are converted into bone, a process called *closure of the epiphyses*
      (2) Once closure occurs, no further increase in bone length is possible
   b. In endochondral formation, bone increases in thickness and is remodeled as it grows
      (1) Addition of newly formed bone from the periosteum increases long bone diameter

(2) Simultaneous resorption of bone near the marrow cavity also increases long bone diameter

c. In intramembranous formation, bone grows by the addition of newly formed bone from the periosteum and simultaneous resorption of bone near the marrow cavity; this process is similar to long bone growth, except that epiphyseal lines do not form

5. Certain substances affect bone formation
   a. Pituitary growth hormone and sex hormones promote bone formation
   b. Bone formation requires sufficient calcium and vitamin D, which promotes calcium absorption and incorporation into the bone matrix

C. Bone resorption

1. Multinucleated cells called *osteoclasts* carry out bone resorption; osteoclasts promote bone remodeling by removing unwanted bone while new bone is forming in other areas
   a. Osteoclasts secrete protein-digesting lysosomal enzymes, which digest the organic bone matrix; they also secrete lactic and citric acids, which dissolve calcium phosphate and other bone minerals
   b. Osteoclasts also actively employ phagocytosis and digest small bone fragments
   c. Calcium and phosphate ions are released into the bloodstream, where they become part of the body's ion pool
2. Adrenal cortical hormones promote bone resorption

D. Joints

1. Joints are the areas where two or more bones meet; they hold the skeleton together
2. The three major types of joints are classified by the degree of movement they permit
   a. *Fibrous joints* (synarthroses) permit no movement
      (1) They unite bones with a thin layer of fibrous connective tissue
      (2) Examples include the sutures between the skull bones
   b. *Cartilaginous joints* (amphiarthroses) allow slight movement
      (1) They unite bones with a fibrocartilage
      (2) Examples include the symphysis pubis, vertebral joints, and sacroiliac joints
   c. *Synovial, or movable, joints* (diarthroses) permit free movement; they are the most common type of joint
      (1) Hyaline cartilage covers the articulating surfaces of the bones in a movable joint
      (2) A strong fibrous capsule surrounds the joint and attaches firmly to both bones; it helps stabilize the joint as do the ligaments, tendons, and muscles that connect the bones of the joint
         (a) Synovial membrane lines the capsule

(b) This membrane secretes a viscous synovial fluid that fills the joint capsule; this fluid lubricates the joint and facilitates movement between articular surfaces

(3) Examples include the shoulder, wrist, hip, knee, and ankle joints

3. Joints also may be classified by their shape or motion, such as ball and socket or hinge and pivot

E. Skeletal tissue

1. *Tendons* are bands of fibrous connective tissue that attach muscles to bones; they enable bones to move when skeletal muscles contract
2. *Ligaments* are dense, strong, flexible bands of fibrous connective tissue that attach bones to bones; they provide stability and may limit or facilitate movement
3. *Cartilage,* which may be fibrous, hyaline, or elastic, is a dense tissue that consists of fibers embedded in a strong, gel-like substance; it supports and shapes various structures, and some cartilage provides a buffer against shock
4. *Bursae* are small synovial fluid sacs that appear around joints between bones, tendons, and ligaments; they decrease stress on nearby tissues by acting as cushions

F. Bone functions

1. The bones protect organs and internal tissues; for example, 33 vertebrae surround and protect the spinal cord
2. The bones support the body; they stabilize it and provide a surface for muscle, tendon, and ligament attachment
3. In concert with the skeletal muscles and joints, the bones allow body movement
   a. Skeletal muscles exert force on tendons, which pull on the bones to which they are attached
   b. During movement, bones act as levers and joints act as fulcrums
   c. The movable joints permit 13 basic types of movement
      (1) *Flexion* decreases the joint angle
      (2) *Extension* increases the joint angle
      (3) *Hyperextension* increases the joint angle beyond the anatomic position
      (4) *Circumduction* moves the limb in a circle
      (5) *Abduction* moves the limb away from midline
      (6) *Adduction* moves the limb toward midline
      (7) *Rotation* revolves the limb around a longitudinal axis, moving it toward midline (internal rotation) or away from midline (external rotation)
      (8) *Supination* turns the palm upward
      (9) *Pronation* turns the palm downward
      (10) *Inversion* turns the plantar surface inward
      (11) *Eversion* turns the plantar surface outward
      (12) *Retraction* moves the jaw backward
      (13) *Protraction* moves the jaw forward
4. The bone marrow produces red blood cells through hematopoiesis (see Chapter 7, Blood, for more information)
5. The bones store minerals; for example, they contain about 99% of the calcium in the body

Study Activities

1. Compare and contrast the three major types of muscles.
2. Discuss the role of acetylcholine in nerve impulse transmission across the neuromuscular junction.
3. Explain the mechanisms by which skeletal muscles contract.
4. Detail the sources of energy for muscle contraction.
5. Explain how cardiac and smooth muscle contractions differ from skeletal muscle contraction.
6. Describe the functions of osteoblasts and osteoclasts.
7. Compare and contrast endochondral and intramembranous bone formation.
8. Discuss the functions of bones and joints.

4

Sensory System

Objectives

After studying this chapter, the reader should be able to:

- Describe the functions of the major structures of the eye.
- Compare the functions and locations of retinal rods and cones.
- Trace the path of a retinal image from the retina through the visual pathways to the brain.
- Explain how the eyes adapt for near vision.
- Define binocular vision and explain how it contributes to depth perception.
- Describe how sound waves are transmitted to the inner ear to produce sound.
- Explain how sound is localized.
- Explain how vestibular structures convey a sense of equilibrium.
- Compare the perceptions of taste and smell.
- Describe how the sense of smell contributes to the sense of taste.
- Locate general sense receptors and explain how they produce sensations of touch, pressure, temperature, and pain.
- Contrast the three types of pain.

I. Sensory System Divisions

A. General information

1. The sensory system may be divided into the special senses —vision, hearing, equilibrium, smell, and taste —and the general senses —touch, pressure, temperature, and pain
2. The special senses are so called because their receptors are found in complex organs in a relatively small area of the body
3. The general senses are so called because their receptors are found throughout the body
4. When stimulated, receptors generate nerve impulses, which are carried to the central nervous system by sensory neurons; the brain then processes this information and makes a suitable response

B. Special senses

1. The senses of vision, hearing, smell, and taste are considered special senses because their receptors are located in complex organs in a relatively small area of the body
    a. Vision receptors exist in the retina of the eye
    b. Hearing receptors exist in the inner ear

c. Smell receptors exist in the upper nose
d. Taste receptors exist in the tongue
e. Receptors for equilibrium exist in the ear

2. Equilibrium also is considered a special sense because it is maintained by the sensory organ of hearing

C. General senses

1. The general senses include touch, pressure, temperature, and pain
2. Receptors for these senses are distributed widely in the skin and other body tissues
   a. Because the number of receptors for each sense varies greatly, not all parts of the body are equally sensitive
   b. The fingertips and tip of the tongue, for example, are much more sensitive than the back of the neck

II. Vision

A. General information

1. The eyes, the sensory organs for vision, collect light waves and transmit them as nerve impulses along the visual pathways to the brain, which translates them into images
2. The eye is a complex structure composed of three layers: an outer cornea and sclera, a middle choroid coat, and an inner retina
3. The cornea and sclera form a dense outer layer, which maintains eye rigidity and protects the inner structures
   a. The cornea is transparent; it allows light to pass through and bends (refracts) the light waves
   b. The sclera is the opaque outer covering
4. The choroid is the vascular, heavily pigmented middle layer
   a. Pigment absorbs the light and prevents it from scattering in the eye
   b. The anterior part of the choroid is modified to form the ciliary body and iris
5. The retina, or inner layer, contains the visual receptors (rods and cones)
   a. Nerve fibers from the retina converge to form the optic nerve
   b. The optic nerve penetrates the posterior surface of the sclera
6. The lens is a transparent, elastic, focusing device enclosed in a capsule and suspended from the ciliary body; lens thickness can be adjusted to focus for near and far vision by contraction or relaxation of the ciliary body, which varies the tension on the suspensory ligaments
7. The space between the cornea and lens is divided by the iris into the anterior and posterior chambers
   a. These chambers are filled *aqueous humor,* a clear watery fluid formed by the ciliary body
   b. Aqueous humor diffuses around the anterior eye structures and is absorbed into a drainage system at the periphery of the anterior chamber in the angle between the cornea and the iris
8. The large space behind the lens is filled with *vitreous humor,* a thick, gelatinous fluid

9. The conjunctiva is the thin vascular membrane that lines the inner surface of the eyelids and sclera; it is attached to the sclerocorneal junction but does not extend over the cornea
10. Lacrimal glands in the orbital cavity discharge fluid secretions to moisten the conjunctiva; the fluid collects in the lacrimal ducts, which have openings on the ridges of the eyelids near the nose, and is discharged into the nose
11. The eyes normally form a clear retinal image of an object 20′ away; they must make several changes to adapt for near vision
12. ***Binocular vision*** contributes to depth perception (ability to judge relative distances of objects)

B. Aqueous humor formation and circulation

1. Aqueous humor is a lymphlike fluid secreted by the ciliary body
2. After the fluid flows into the posterior chamber of the eye, it flows into the anterior chamber by passing through the pupil; it is absorbed into the *canal of Schlemm,* a ring-shaped canal that encircles the eyeball at the sclerocorneal junction
3. The portion of the eyeball between the anterior chamber and the canal of Schlemm is porous and contains a network of channels, called the *trabecular meshwork,* that leads to the canal
   a. Aqueous humor flows through the trabecular meshwork into the canal of Schlemm
   b. Then the fluid enters the veins that drain blood from the eyeball
4. Normally, aqueous humor secretion balances its absorption, and the pressure in the anterior chamber remains relatively constant at about 20 to 25 mm Hg (about one-fifth of systemic arterial pressure); elevated intraocular pressure may lead to vision loss when the pressure is transmitted to the vitreous humor, where it damages retinal neurons

C. Light refraction

1. Light rays that enter a transparent medium at an oblique angle to the surface are refracted as they pass into another medium of different density, such as from air to glass or from air to the eye
2. The degree of ***refraction*** depends on the angle of the light rays and the difference in the refractive indices (measures of the density of the media) of the two media through which light rays pass
   a. The more oblique the angle of the light rays, the greater the refraction
   b. The greater the difference in the refractive indices, the greater the refraction; for example, light rays passing from air (with a ***refractive index*** of 1) to glass (with a refractive index of 1.4) are bent more than rays passing from water (refractive index of 1.33) to glass because the difference between the refractive indices of air and glass is greater than that between the indices of water and glass
3. The cornea is an efficient refracting medium because its refractive index (1.33) is significantly higher than that of air (1), which the light rays travel through before striking the cornea
4. Light rays also can be refracted by lens curvature; the greater the curvature, the more the rays are refracted
   a. Concave lenses (those curved inward) cause light rays to diverge

b. Convex lenses (those curved outward), such as the cornea and lens of the eye, cause light rays to converge
5. Parallel rays from an object more than 20′ away that strike a convex lens come into focus at the ***focal point*** *(principal focus);* the distance from the lens to its focal point is the ***focal length*** of the lens
6. The converging power of a convex lens is expressed in *diopters,* the reciprocal of the focal length of the lens expressed in meters
   a. A lens with a focal length of 1 meter has a strength of 1 diopter
   b. A lens with a focal length of 50 cm (½ meter) has a strength of 2 diopters because the reciprocal of ½ is 2⁄1, or 2
7. The refractive power of a concave lens cannot be expressed in focal length because this lens causes light rays to diverge instead of converge at a focal point
   a. Refraction power of a concave lens is determined by its ability to counteract the converging power of a convex lens; its power is expressed by a minus sign
   b. A concave lens that counteracts the converging power of a 0.5-diopter lens has a refractive power of –0.5 diopters

D. Image formation on the retina

1. The eye has a refractive power of about 60 diopters
   a. The cornea performs most of the refraction because it is a convex lens and has a high refractive index (1.33)
   b. The lens has much less refractive power than the cornea because it is surrounded by media (aqueous and vitreous humors) that have almost the same refractive index as the cornea
   c. However, the lens can change its focal length and bring images into sharp focus on the retina through ***accommodation***
2. After passing through the cornea, aqueous humor, lens, and vitreous humor, light waves hit the retina
3. Receptor cells of the retina (rods and cones) contain a photosensitive visual pigment (rhodopsin) that decomposes on light exposure, stimulating an impulse that is conveyed to the brain
   a. Rods are the most numerous photoreceptors
      (1) They are concentrated at the periphery of the retina
      (2) They are sensitive to low levels of illumination but cannot discriminate color
   b. Rods do not function in bright light because it decomposes most of their rhodopsin
      (1) This explains why a person cannot see well for a short time after going from bright light into darkness
      (2) After a brief period of darkness, rhodopsin reforms and the rods start functioning again
   c. Cones are less sensitive to light
      (1) They are concentrated in the center of the retina
      (2) They provide daylight color vision
   d. Cones come in three types, responding to red, green, or blue light
      (1) Stimulation of various receptor combinations transmits impulses that the brain interprets as color

(2) Color blindness occurs when one or more types of cones are absent or defective

4. Images form in the eye similar to the way they form in an autofocus camera
   a. The image on the retina is inverted
   b. The brain reverses the image and perceives the object right side up

E. Visual pathways

1. An image forms on the retina when light stimulates the rods and cones
2. Each half of the retina receives visual input from the opposite visual field
   a. The right half of each retina receives input from the left visual field
   b. The left half of each retina receives input from the right visual field
3. Two optic nerves converge at the base of the brain to form the *optic chiasma*
   a. The nerve fibers continue behind the chiasma, forming the ***optic tracts***
   b. These tracts connect to neurons in the brain stem, which convey visual impulses to the occipital cortex by means of fiber tracts called the *optic radiations*
4. Nerve fibers from the medial halves of both optic nerves cross in the optic chiasma, and fibers from the lateral halves of the optic nerves continue into the optic tracts without crossing
5. Each half of the retina receives an image of objects from the visual field on the opposite side
   a. The right optic tract contains nerve fibers from the right half of each retina
   b. The left optic tract contains nerve fibers from the left half of each retina

F. Near vision

1. In an eye with normal refractive power, the retina forms a clear image of an object 20′ away
2. Viewing of objects closer than 20′ from the eye (near vision) requires three automatic changes: accommodation, pupillary constriction, and eye convergence
3. *Accommodation* (adjustment of the refractory power of the lens) depends on the elasticity of the lens, which is surrounded by a strong capsule attached to suspensory ligaments
   a. When the eye is at rest, suspensory ligaments hold the lens under tension, compressing and flattening it
   b. During accommodation, the ciliary muscles contract, pulling the suspensory ligaments forward and relaxing their tension on the lens capsule; this makes the lens bulge and become more convex
   c. Accommodation increases the refractive power of the eye, which is necessary for near vision
4. *Pupillary constriction* blocks light rays that normally pass through the lens periphery; such rays are refracted more than those passing through the center because the refractive index at the periphery (1.36) is slightly different than that at the center (1.42)
   a. Blocking these rays increases the sharpness of the image because all the rays are focused more sharply on the retina
   b. The pupils also constrict in response to bright light to protect the eyes
5. ***Convergence*** allows the individual to see an object close up with both eyes without seeing double because the two images are focused on corresponding points on the two retinas

G. Binocular vision

1. *Binocular vision* (single vision with two eyes) permits depth or distance perception and creates a larger field of vision
2. During convergence, when the image falls at corresponding points on both retinas, each eye views the object from a slightly different angle
3. The two slightly different images are relayed to the brain, where they are synthesized (fused) into a single image, which conveys an impression of depth that the separate retinal images lack

III. Hearing

A. General information

1. The ears, the sensory organs for hearing, gather sound waves and transmit them as nerve impulses to the brain
2. The brain interprets these impulses as hearing
3. The auditory apparatus consists of the external ear, middle ear, and inner ear
4. The external ear collects sound; it is composed of a fibrous and cartilaginous framework covered by skin that is continuous with the skin covering the ear canal (external auditory canal)
5. The middle ear conducts sound; it consists of a small cavity that contains the ossicles (three small bones —malleus, incus, and stapes —that transmit sound to the inner ear) and is vented by the eustachian tube
   a. Two small muscles attached to the ossicles contract automatically in response to loud noises
   b. This contraction dampens ossicle vibrations, which normally are transmitted to the inner ear; thus, it protects inner ear structures from damage
   c. The middle ear is separated from the ear canal and external ear by the tympanic membrane and from the inner ear by two openings (the oval window and the round window)
6. The inner ear lies in a cavity in the temporal bone (bony labyrinth); it is a collection of tubes and sacs (membranous labyrinth)
7. The *bony labyrinth* consists of three parts: the bony semicircular canals, vestibule, and cochlea
8. The *membranous labyrinth* consists of four parts enclosed within the bony labyrinth: the membranous semicircular canals inside the bony semicircular canals, two chambers within the vestibule called the utricle and saccule, and the cochlear duct within the cochlea
   a. The *semicircular canals* open into the utricle
   b. The *utricle* is connected to the saccule
   c. The *saccule* is connected to the cochlear duct
9. The membranous labyrinth is filled with a fluid called *endolymph* and surrounded by a fluid called *perilymph,* which fills the bony labyrinth
10. The *cochlea* (named for its resemblance to a snail shell) is the part of the bony labyrinth that contains the cochlear duct
    a. The cochlea consists of a spiral bony tube that makes several turns around a central bony column, similar to a spiral staircase surrounding a central pillar
    b. It contains three channels that extend its entire length

(1) One channel is the endolymph-filled cochlear duct, which is part of the membranous labyrinth and contains the organ of hearing *(organ of Corti)*

(2) The vestibular and tympanic canals are the other channels, which are located on opposite sides of the cochlear duct and filled with perilymph; they are connected by a narrow opening called the *helicotrema* at the apex of the cochlea

(3) The vestibular canal also connects with the middle ear by the oval window, which is covered by a membrane and by the footplate of the stapes

(4) The tympanic canal connects with the middle ear by the round window, which is covered by a flexible membrane

(5) Vestibular membrane lines the portion of the cochlear duct wall next to the vestibular canal; basilar membrane lines the portion of the wall next to the tympanic canal

11. The organ of Corti, which extends the length of the cochlear duct, is composed of auditory receptor cells (hair cells), supporting cells, and nerve fibers
    a. Receptor cells are embedded in the basilar membrane of the cochlear duct; their free surfaces project into the endolymph of the duct
    b. A gelatinous membrane called the *tectorial membrane* overhangs and touches these hair cells
    c. Stimulation of these cells causes sound wave transmission

B. Sound wave transmission

1. Hearing requires sound wave reception and conduction to the organ of Corti, where the waves are converted into nerve impulses; then the nerve impulses are transmitted to the auditory area in the cerebral cortex
2. Sound waves may arise from any vibration source, such as the vocal cords or a musical instrument
3. Sound waves differ in pitch, which reflects the number of vibrations (cycles) per second (CPS); the higher the pitch, the greater the number of CPS
    a. The human ear can detect tones as low as 30 CPS and as high as 20,000 CPS
    b. It is most sensitive to sounds that range from 500 to 4,000 CPS
4. Sound waves may be transmitted to the inner ear by vibrations of the tympanic membrane and ossicles ***(air conduction)*** or by vibrations of the skull, which bypass the tympanic membrane and ossicles ***(bone conduction)***
5. In air conduction, the external ear funnels sound waves into the ear canal, where they strike the tympanic membrane, causing it to vibrate
    a. These vibrations are transmitted through the middle ear to the oval window, which is covered by the footplate of the stapes
    b. Because the surface area of the tympanic membrane is 20 times greater than that of the oval window, sound waves are concentrated and amplified at the footplate
    c. Footplate vibrations are transmitted to the perilymph of the vestibular canal and then, via the helicotrema, into the perilymph of the tympanic canal
    d. Each inward movement of the oval window causes a corresponding outward movement of the round window at the base of the tympanic canal, causing waves in the perilymph

e. Perilymph vibrations in the vestibular and tympanic canals set up corresponding vibrations in the basilar membrane of the cochlear duct
   (1) Basilar membrane vibrations cause the processes of the hair cells in contact with the tectorial membrane to bend, which stimulates these cells; the cells stimulate neurons that synapse on the hair cells
   (2) Sounds of different frequencies cause vibrations in different parts of the basilar membrane, which stimulate hair cells in different parts of the organ of Corti

f. The cochlear branch of the vestibulocochlear nerve (cranial nerve VIII) collects nerve impulses initiated by stimulated hair cells and transmits them to the brain

6. In bone conduction, the sound waves pass through the bones of the skull and cause perilymph and basilar membrane vibrations, which activate the hair cells in the same way that they are activated in air conduction

C. Sound perception and localization

1. The auditory center in the cerebral cortex interprets the pitch of a sound based on the part of the organ of Corti that is stimulated by vibrations
2. Differences in loudness result from variations in hair cell stimulation; loud sounds stimulate more hair cells, which generate more impulses than soft sounds
3. The brain interprets the direction of sound based on slight differences in the arrival time and intensity of the sound in each ear

IV. Equilibrium

A. General information

1. The sense of ***equilibrium,*** or balance, is vital for maintaining stability during movement or at rest
2. The vestibular apparatus of the inner ear controls the sense of equilibrium and position

B. Vestibular apparatus structure and function

1. The vestibular apparatus consists of the membranous semicircular canals, which respond to motion, and the utricle and the saccule, which respond to position changes
2. The three tubes of the semicircular canals are connected to the utricle; each tube lies at right angles to the other two
   a. One end of each semicircular canal expands slightly where it joins the utricle and contains groups of specialized receptors called *hair cells,* as well as supporting cells
      (1) These hair cells are similar to those in the organ of Corti
      (2) They are covered by gelatinous material called *cupula*
      (3) They are stimulated by endolymph movement in the semicircular canals during body movement
   b. These hair cells transmit nerve impulses over the vestibular branch of the vestibulocochlear nerve to the medulla, where they are interpreted as motion; impulses also are relayed to the cerebellum, where they initiate reflex movements of body muscles and eye muscles

3. The utricle and saccule convey a sense of head position to the brain by sending impulses to the cerebellum through the vestibulocochlear nerve
   a. The utricle and saccule are filled with endolymph and contain hair cells; the free ends of the hair cells are embedded in a mass of gelatinous material that contains calcium carbonate crystals called *otoliths*
   b. Gravity presses on the otoliths and the cilia of the hair cells, stimulating nerve impulses that the brain interprets as the normal head position
   c. During head movement, the weight of the otoliths shifts and the cilia bend, transmitting nerve impulses that are perceived by the brain as a change in head position
   d. The impulses also initiate automatic reflex reactions that restore the head to its normal position

V. Smell

A. General information

1. Olfactory receptors are specialized neurons with dendrites that are modified to respond to odors
2. These receptors are located in the roof of the nasal cavity; they relay impulses to the olfactory bulbs in the cranial cavity, which transmit impulses to cortical neurons where odor is perceived

B. Sensation of smell

1. Air must pass through the nose to stimulate the olfactory receptors
2. A person cannot smell if the nostrils are blocked because no air can enter the nose to stimulate the olfactory receptors
3. Olfactory receptors are sensitive to very low concentrations of odors, but they rapidly adapt to odors
   a. After smelling an odor for a short time, the individual no longer perceives it as intensely
   b. This loss of odor sensitivity results from reduced responsiveness of the olfactory receptors and diminished perception of the odor in the olfactory portion of the cerebral cortex
   c. However, the cerebral cortex stores memories of odors; after smelling an odor once, an individual can recognize it readily if smelled again

VI. Taste

A. General information

1. Taste receptors are located in the taste buds, which are concentrated in the tongue but also exist in the soft palate and throat
2. Four types of taste receptors exist; each type can sense one of four basic tastes: sweet, sour, bitter, and salt
   a. Sweet tastes are perceived on the tip of the tongue
   b. Sour tastes are perceived along the sides of the tongue
   c. Bitter tastes are perceived on the back of the tongue
   d. Salty tastes are perceived on the tip and sides of the tongue

B. Sensation of taste

1. For taste to occur, substances must be in solution in saliva
2. Taste perception results from taste bud stimulation as well as olfactory receptor stimulation from air passage through the nose
3. Smell, texture, and temperature contribute to the overall perception of taste, for example, of food flavors
4. If the sense of smell is not functioning properly (for example, if the nose is congested from a cold), the taste perception may be diminished or unusual

VII. General Senses

A. General information

1. The general senses include touch, pressure, temperature, and ***pain***
2. Receptors for these senses are distributed widely throughout the skin and other body tissues
3. Because the number of receptors for each general sense varies widely, body parts are not equally sensitive to stimulation

B. Touch and pressure

1. Receptors for these sensations lie in nerve endings around hair follicles and in the papillary layer of the skin
2. When stimulated, these receptors transmit a nerve impulse that is carried by a cranial nerve to the brain or by a spinal nerve to the spinal cord and then to the brain
3. In the brain, the general sensory area located behind the central fissure interprets these impulses as touch or pressure

C. Temperature

1. Cold receptors lie near the surface of the skin
2. Heat receptors lie deep in the skin
3. When stimulated these receptors transmit a nerve impulse via the same transmission route as that of touch and pressure receptors
4. These impulses are perceived and interpreted in a similar manner to impulses stimulated by touch and pressure

D. Pain

1. Pain receptors are located throughout the body in the skin, muscles, tendons, and joints
2. Pain serves a protective function by alerting the individual to withdraw from a harmful stimulus; loss of the ability to feel pain makes the individual vulnerable to injury
3. Pain impulses are transmitted by myelinated and unmyelinated nerve fibers via the same transmission route as the other general senses
    a. Impulses conducted rapidly by myelinated fibers are perceived as sharp pain
    b. Impulses conducted more slowly by unmyelinated fibers are perceived as dull, aching pain
4. Different types of pain result from stimulation of different areas
    a. Pain in the skin, subcutaneous tissues, muscles, bones, and joints is called *somatic pain*

b. Pain in the internal organs (from distention, smooth muscle spasm, or inadequate blood supply) is called *visceral pain*

c. Internal organ pain that seems to come from the body surface at a distant site is called ***referred pain;*** for example, the pain of a myocardial infarction may be felt in the neck or along the inner arm

(1) Referred pain results because pain impulses from receptors in an internal organ enter the same part of the spinal cord as impulses from somatic pain receptors on the body's surface

(2) The pain impulses are conveyed along pain pathways in the spinal cord to the brain

(3) The brain misinterprets these sensations as coming from the pain receptors on the body surface rather than from those in the internal organ

Study Activities

1. Differentiate the special senses from the general senses.
2. Discuss the major structures of the eye and retina.
3. Identify the visual pathways from the eye to the brain.
4. Compare near vision and binocular vision.
5. Describe the functions of the major structures of the ear.
6. Contrast air conduction and bone conduction of sound.
7. Explain how the body maintains a sense of equilibrium.
8. Describe how the senses of smell and taste are related.
9. Identify the four general senses and explain how they are interpreted in the brain.

5

Respiratory System

Objectives

After studying this chapter, the reader should be able to:

- Describe the basic structure of the lungs and their function in respiration and gas exchange.
- Explain how surfactant prevents the alveoli from collapsing during respiration.
- Discuss the importance of negative intrapleural pressure.
- Describe how the diaphragm and intercostal muscles help move air in and out of the lungs.
- Contrast the various pulmonary volumes and capacities.
- Explain how the partial pressure of a gas relates to its concentration in a gas mixture.
- Compare the concentrations of oxygen and carbon dioxide in inspired (atmospheric), alveolar, and expired air.
- Explain how oxygen is transferred from the lungs to the tissues and how carbon dioxide is transferred from the tissues to the lungs.
- Describe how oxygen and carbon dioxide are transported in the body.
- Explain how the respiratory center in the brain controls respiration.

I. Respiratory Structures and Functions

A. General information

1. The respiratory system, which consists of respiratory passages and lungs, works with the cardiovascular system to oxygenate blood and eliminate carbon dioxide
2. *Respiration* consists of two processes
   a. ***Ventilation*** refers to air movement through the respiratory passages to and from the lungs
   b. ***Gas exchange*** refers to oxygen and carbon dioxide transport between the pulmonary alveoli and the blood in the pulmonary capillaries
   c. Both processes must function properly to achieve adequate tissue oxygenation and efficient carbon dioxide elimination

B. Respiratory passages

1. Respiratory passages, which are responsible for ventilation, consist of upper and lower airways

a. *Upper airways* include the nasal cavity, pharynx, and larynx
b. *Lower airways* include the trachea, bronchi, and bronchioles
2. The *nasal cavity* filters, warms, and moistens air inspired through the nostrils; it connects to the pharynx
3. The tubelike *pharynx* is a common pathway for the respiratory and gastrointestinal systems; it extends down from the juncture of the nasal and oral cavities and splits into the larynx and the esophagus
4. The *larynx* is a boxlike structure that contains the vocal cords and connects the pharynx to the trachea; vocal cords produce sounds by vibrating when expired air passes across their surfaces
5. The *trachea, bronchi,* and *bronchioles* are flexible tubes; air travels through them to gas-exchange sites in the lungs
   a. The trachea extends down from the larynx; it divides into the right and left primary bronchi, each of which extends into one lung
   b. Bronchi branch repeatedly, becoming progressively smaller; eventually, they form tiny bronchioles (the first respiratory passages that contain no cartilage in their walls)
   c. Bronchioles also branch repeatedly, forming terminal bronchioles (smallest bronchioles that conduct air only) and respiratory bronchioles (tubes distal to the terminal bronchioles that conduct air and participate in gas exchange)
   d. The diameter of the bronchi and bronchioles varies with the respiratory phase; they increase slightly during inspiration and decrease slightly during expiration

C. Lungs

1. The lungs are air-filled, sponge-like organs that contain *pulmonary alveoli,* grapelike clusters at the end of respiratory passages; gas exchange takes place in the alveoli by diffusion
   a. Alveoli are separated by thin vascular partitions called alveolar septa, which are lined by flat squamous cells and small numbers of secretory cells
   b. The secretory cells produce *surfactant*
      (1) The lipoprotein surfactant reduces surface tension, allowing the fluid that lines the alveoli to spread as a thin film rather than coalescing into droplets
      (2) Surfactant reduces the cohesive surface tension of water molecules in the alveoli, allowing the alveoli to expand uniformly during inspiration; without surfactant, surface tension could restrict alveolar expansion or cause alveolar collapse during expiration
2. The left and right lungs are separated by the mediastinum, which contains the heart, blood vessels, and other midline structures; fissures divide each lung into lobes
3. The lungs and pleural cavities are covered by a thin serous membrane called the pleura
   a. The visceral pleura covers the lungs and dips into the fissues
   b. At the hilus, (an indentation in an organ where nerves and blood vessels enter), the visceral pleura folds back to form the parietal pleura, which lines the chest wall and covers the diaphragm
   c. The space between the lungs and chest wall is called the pleural cavity

d. When fully expanded, the lungs completely fill the pleural cavity, and the parietal and visceral pleurae come in contact

4. The lungs remain expanded in the pleural cavity because the ***intrapleural pressure*** (pressure in the pleural cavity) is less than the ***intrapulmonary pressure*** (air pressure in the lungs)
5. These pressure differences develop at birth when the thoracic cavity enlarges and respiration begins
6. As the lungs expand and fill with air at atmospheric pressure, the elastic tissue of the lungs stretches
7. The stretched lungs tend to pull away from the chest wall and return to their original state; this creates a slight vacuum in the pleural cavity and makes the intrapleural pressure slightly less than atmospheric pressure
8. Because the intrapleural pressure is slightly less than atmospheric pressure, it commonly is called negative or subatmospheric pressure; without this pressure to hold the lung in the expanded position, the elastic tissue would contract and the lungs would collapse

II. Ventilation

A. General information

1. The diaphragm and intercostal muscles produce the normal inspiratory and expiratory movement of the lungs and ribs; this movement allows ventilation (air movement in and out of the lungs)
2. During respiration, the intrapulmonary and intrapleural pressures change from their levels at rest, which are 760 and 756 mm Hg, respectively; the atmospheric pressure remains constant at about 760 mm Hg
3. The volume of air in the lungs varies during the phases of respiration, *inspiration* (air movement into the lungs) and *expiration* (air movement out of the lungs)
4. Pulmonary volume (amount of air in the lungs) varies greatly between normal inspiration and expiration; it varies even more with forced inspiration and expiration

B. Inspiration

1. Stimulated by the central nervous system, the diaphragm contracts and descends, pulling down the lower surfaces of the lungs
2. At the same time, the external intercostal muscles contract and raise the rib cage; this expands the lungs by lifting the sternum up and forward
3. Thoracic expansion lowers the intrapleural pressure to 754 mm Hg, and the lungs expand to fill the enlarged thoracic cavitiy
4. The intrapulmonary pressure decreases to 758 mm Hg; the intrapulmonary–atmospheric pressure gradient pulls air into the lungs

C. Expiration

1. Normally, expiration is a passive process
2. As central nervous system impulses cease after inspiration, the diaphragm slowly relaxes and moves up

3. The external intercostal muscles relax and the rib cage descends
4. These actions allow the lungs and thorax to return to their resting size and position
5. Lung and thorax relaxation causes the intrapulmonary pressure to rise above the atmospheric pressure to 763 mm Hg; the intrapleural pressure rises to 756 mm Hg
6. The intrapulmonary–atmospheric pressure gradient forces air out of the lungs until the two pressures are equal
7. During vigorous exertion, the lungs can expel air more actively
   a. Contraction of the internal intercostal muscles actively forces down the ribs
   b. Abdominal muscle contraction pushes the abdominal contents up against the bottom of the diaphragm, expelling air more rapidly than is possible with normal diaphragm and intercostal muscle relaxation

D. Pulmonary volumes and capacities

1. Pulmonary volumes and capacities vary with the individual's size, which influences the size of the lungs and thorax
2. During normal quiet breathing, the average adult inspires and expires about 500 ml of air with each breath, which is called the ***tidal volume***
3. At the end of a normal tidal inspiration, an adult can forcefully inhale an additional 3,000 ml of air, which is called the ***inspiratory reserve volume***
4. At the end of a normal tidal expiration, an adult can forcefully exhale about 1,300 ml of air, which is called the ***expiratory reserve volume***
5. Even if the expiratory reserve volume is expelled from the lungs, an additional 1,200 ml of air remains; this remaining air is called the ***residual volume***
6. The maximum amount of air that can be moved out of the lungs after a maximum inspiration and expiration, called the ***vital capacity,*** is about 4,800 ml (500 ml + 3,000 ml + 1,300 ml)
7. The *total lung capacity* is approximately 6,000 ml
8. Although the tidal volume is 500 ml, about 150 ml of this air never reaches the alveoli; it fills the upper respiratory passages and is exhaled with the next breath
   a. The upper respiratory passages containing air that does not enter the alveoli is called the *dead space;* the volume of air it contains is called the *dead space volume*
   b. Because of this dead space, only about 350 ml of air actually enters the alveoli with each breath and mixes with the 2,500 ml of air already in the lungs (expiratory reserve volume plus the residual volume)
   c. Consequently, less than 15% of alveolar air is replaced by new atmospheric air with each breath
   d. This slow replacement of alveolar air has several advantages
      (1) The slow admixture of atmospheric air with alveolar air prevents wide fluctuations in oxygen and carbon dioxide concentration in alveolar air, which could produce adverse effects
      (2) Because oxygen and carbon dioxide in the blood are in equilibrium with their gas forms in the alveolar air, stable concentrations of alveolar oxygen and carbon dioxide promote stable concentrations of these gases in the arterial blood

III. Principles of Gas Exchange

A. General information

1. Oxygen and carbon dioxide diffusion (exchange) between the alveoli, blood, and tissues depends on the concentrations and pressures of these gases
2. In a mixture of gases, the pressure exerted by each gas ***(partial pressure)*** is independent of that of the other gases and directly corresponds to the percentage (concentration) it represents of the total mixture
   a. For example, air exerts a total pressure of 760 mm Hg (atmospheric pressure) at sea level
   b. Air contains about 21% oxygen; consequently, the partial pressure of oxygen (PO_2) is 21% of the atmospheric pressure: $0.21 \times 760 = 159.6$ mm Hg
3. Oxygen and carbon dioxide exist in three physiologically important areas: in the atmosphere and pulmonary alveoli as gases, and in the blood in solution

B. Gas concentrations in inspired air

1. Most air inspired from the atmosphere (almost 79%) consists of the inert gas nitrogen; air contains only about 21% oxygen
2. The rest is a mixture of small amounts of water vapor (about 0.5%) and carbon dioxide (about 0.04%)

C. Gas concentrations in alveolar air

1. Alveolar air contains more water vapor (about 6.2%) than inspired air because of the moist secretions in the respiratory passages
2. The oxygen content (about 13.6%) is lower than that of inspired air because erythrocytes take up oxygen as they pass through the pulmonary capillaries
3. The carbon dioxide concentration (about 5.3%) is higher than that of inspired air because the gas diffuses continually from the pulmonary capillaries into the alveolar air
4. Nitrogen accounts for about 74.9% of alveolar air

D. Gas concentrations in expired air

1. Expired air contains about the same amount of water vapor as alveolar air (about 6.2%)
2. Expired air contains more oxygen (about 15.7%) and less carbon dioxide (about 3.6%) than alveolar air because it is a mixture of alveolar air and atmospheric air from the dead space
3. Nitrogen accounts for about 74.5% of expired air

E. Gas diffusion (exchange)

1. *Diffusion* refers to the exchange of gases —particularly oxygen and carbon dioxide —between the alveoli and capillaries and between body cells and erythrocytes
2. In diffusion, substances move from an area of higher concentration to one of lower concentration; a gas diffuses from an area with a high partial pressure of the gas to one with a low partial pressure
3. Inspired air has a PO_2 of 158 mm Hg and a PCO_2 of 0.3 mm Hg

4. When blood returns to the heart, the right ventricle pumps it to the lungs, where it passes through the pulmonary capillaries
5. The PO_2 in this blood, which is returned to the heart by the vena cavae and pumped to the lungs is 40 mm Hg; the PCO_2 is 47 mm Hg
6. Because alveolar air has a higher PO_2 (100 mm Hg) and a lower PCO_2 (40 mm Hg) than the blood in the pulmonary capillaries, oxygen diffuses from alveolar air into the pulmonary capillaries, and carbon dioxide diffuses in the opposite direction
7. As a result, arterial blood has a PO_2 of 97 mm Hg, which is almost the same as the alveolar PO_2; it also has a PCO_2 of 40 mm Hg, which is the same as the alveolar PCO_2
8. In the tissues, the PO_2 is lower (40 mm Hg) and the PCO_2 is higher (about 60 mm Hg) than in arterial blood
9. Therefore, oxygen diffuses into the tissues from the arterial blood, and carbon dioxide diffuses in the opposite direction, from the tissues into the blood

IV. Oxygen and Carbon Dioxide Transport

A. General information

1. Once diffusion occurs, arterial blood transports oxygen to the tissues in two ways: physically dissolved in plasma and chemically bound to hemoglobin
2. The tissues release carbon dioxide into the bloodstream, where it travels to the lungs in three forms: dissolved in plasma, combined with hemoglobin, and combined with water as carbonic acid and its component ions
3. In the lungs, oxygen and carbon dioxide transport is the reverse of that in the tissues

B. Oxygen transport in blood

1. Only about 3% of oxygen is dissolved in blood plasma
2. The remaining 97% is chemically bound with hemoglobin in a ratio of 1 g of hemoglobin to approximately 1.34 cubic centimeters (cc) of oxygen
   a. 100 ml of blood contains about 15 g of hemoglobin; therefore, 100 ml of blood should combine with about 20 cc of oxygen ($15 \times 1.34 = 20.1$)
   b. However, oxygenated blood normally carries slightly less than the predicted 20 cc of oxygen because it is only about 97% oxygen-saturated; in other words, the hemaglobin does not carry its full complement of oxygen
3. The tissues remove about 5 cc of oxygen from each 100 ml of blood; blood returning to the heart contains about 15 cc of oxygen per 100 ml of blood (75% oxygen saturation)
4. During vigorous exercise, muscles remove more oxygen from the blood, and the rate of blood flow to the tissues increases greatly, which increases the amount of oxygen available to the tissues
5. Oxygen uptake by hemoglobin is most efficient in the lungs, where the oxygen concentration is high; oxygen release occurs most readily in the tissues, where the oxygen concentration is low

6. The lower acidity (pH) and higher temperature of actively metabolizing tissue, which may occur during vigorous exercise, also enhances oxygen release from hemoglobin

C. Carbon dioxide transport

1. Carbon dioxide is transported in several forms in the blood
   a. A small amount is dissolved in plasma
   b. Some is loosely combined with amino groups in the hemoglobin molecule
   c. Most of the carbon dioxide is converted in erythrocytes to bicarbonate by the enzyme carbonic anhydrase
      (1) Conversion occurs when oxygen is liberated from hemoglobin
      (2) Bicarbonate combines with sodium and is transported in the plasma as sodium bicarbonate
2. Erythrocyte uptake of carbon dioxide begins in the capillaries
   a. In the erythrocytes, hemoglobin liberates oxygen to supply the tissues
   b. Simultaneously, carbon dioxide diffuses from the tissues into erythrocytes, where carbonic anhydrase rapidly catalyzes carbonic acid formation by combining with carbon dioxide and water
   c. Then carbonic acid dissociates into hydrogen ions (H^+) and bicarbonate ions (HCO_3^-)
   d. During oxygen release to the tissues, hemoglobin molecules take up hydrogen ions; after releasing oxygen, the molecule is reduced and has an increased capacity to combine with hydrogen ions
   e. The remaining bicarbonate ions accumulate until their concentration in erythrocytes exceeds that in plasma; then bicarbonate ions diffuse from erythrocytes into plasma
   f. Simultaneously, chloride ions diffuse into erythrocytes to replace the bicarbonate ions; this exchange is called the *chloride shift*
   g. Then the bicarbonate ions combine with plasma sodium ions to form sodium bicarbonate

D. Gas transport in the lungs

1. In the lungs, the carbon dioxide–oxygen exchanges are the reverse of those in the tissues
2. Reduced hemoglobin takes up oxygen, decreasing the hemoglobin's capacity to combine with hydrogen ions
   a. Then hemoglobin releases hydrogen ions
   b. Simultaneously, bicarbonate ions diffuse into erythrocytes and chloride ions diffuse out
   c. As bicarbonate ions move into erythrocytes, they combine with hydrogen ions liberated from hemoglobin to form carbonic acid
   d. Carbonic acid rapidly decomposes, liberating carbon dioxide
   e. Carbon dioxide diffuses from erythrocytes into plasma
   f. Some carbon dioxide travels to the alveoli and is excreted by the lungs
   g. Most remains in plasma as part of bicarbonate ions, which play an essential role in maintaining the ***acid-base balance*** of the blood

V. Control of Respiration

A. General information

1. A control center in the brain stem, called the ***respiratory center,*** regulates the rate and depth of respiration
2. This center discharges impulses to neurons that innervate the diaphragm and intercostal muscles
3. The impulse discharge rate from the respiratory center is influenced by the chemical composition of arterial blood and by nerve impulses sent to the respiratory center

B. Arterial blood

1. Neurons in the respiratory center are stimulated directly by increased arterial concentrations of carbon dioxide and hydrogen ions
2. Normally, the respiratory center is regulated primarily by arterial PCO_2; it adjusts respirations automatically to variations in this level
3. During exercise, more carbon dioxide is produced and the PCO_2 rises
   a. This stimulates the respiratory center to increase respiratory rate and depth
   b. This eliminates carbon dioxide more rapidly by the lungs, allowing the PCO_2 to decrease to normal
4. Chemoreceptors in the aortic arch and carotid sinus also convey impulses to the respiratory center
5. These chemoreceptors respond to arterial blood changes in PCO_2, hydrogen ion concentration (acidity), and PO_2; they signal the respiratory center to adjust the respiratory rate and depth as needed

C. Nerve impulses

1. Nerve impulses from the lungs, cerebral cortex, and sensory nerve endings affect the rate of impulse discharge from the respiratory center
2. Receptors in the lungs respond to stretching as the lungs inflate, sending impulses to the respiratory center to inhibit further inspiration
   a. When the lungs deflate (expiration), the stretch receptors no longer are stimulated
   b. Inhibitory impulses no longer are transmitted to the respiratory center
   c. Inspiration follows automatically
   d. This pulmonary reflex mechanism prevents lung overinflation and helps maintain normal respiratory rhythm
3. The cerebral cortex sends impulses to the respiratory center in response to strong emotions, such as anxiety, fear, and anger; these impulses increase the respiratory rate
4. Some sensory stimuli, such as irritating vapors in the upper respiratory passages, may cause reflex inhibition of respiration

Study Activities

1. Identify the components of the respiratory system.
2. Explain the mechanics of ventilation, noting intrapulmonary and intrapleural pressure changes.
3. Describe the common pulmonary volumes and capacities.
4. Explain the basic priciples of gas exchange, comparing gas concentrations in inspired, alveolar, and expired air.
5. Discuss the exchange of oxygen and carbon dioxide between the alveoli and capillaries and between body cells and erythrocytes.
6. Describe how arterial blood and nerve impulses can affect the respiratory center.

6

Cardiovascular System

Objectives

After studying this chapter, the reader should be able to:

- Describe the roles of various structures in the major functions of the cardiovascular system.
- Compare and contrast the pulmonary and systemic circulatory systems.
- Describe the events in the cardiac cycle.
- Explain the cause of normal heart sounds and the pulse.
- Trace the normal cardiac conduction route.
- Define blood pressure and compare its systolic and diastolic components.
- Explain how changes in cardiac output and peripheral resistance can affect blood pressure.
- Explain how Starling's law, baroreceptors, chemoreceptors, and hormones help regulate cardiac output and blood pressure.
- Describe four factors that affect fluid flow between the capillaries and interstitial tissues.

I. Cardiovascular Structures and Functions

A. General information

1. The *cardiovascular system* moves blood throughout the body; it helps maintain proper body pH and electrolyte composition and helps regulate body temperature
2. This system consists of the *blood vessels* and *heart*
    a. Blood vessels include *arteries, arterioles, capillaries, venules,* and *veins*
    b. The heart pumps the blood through the blood vessels

B. Blood vessels

1. A vast network of vessels supplies blood to and from every functioning cell in the body; this network has two branches
    a. The *pulmonary circulation* carries blood to the lungs for oxygenation and for carbon dioxide removal
    b. The *systemic circulation* delivers blood with oxygen and nutrients to cells throughout the body and transports wastes to the kidneys, liver, and skin for excretion
2. Five basic types of blood vessels exist: arteries, arterioles, capillaries, venules, and veins

a. Arteries carry blood away from the heart; they connect the heart to the arterioles
   (1) Arterial walls typically consist of an outer coat (tunica adventitia), middle coat (tunica media), and an inner coat (tunica intima)
   (2) In arteries, vascular resistance to blood flow is low; mean arterial pressure (average of diastolic and systolic pressures) remains around 100 mm Hg
b. Arterioles lie between arteries and capillaries
   (1) Arterioles consist of the same three coats as arteries, but are smaller in diameter
   (2) Vascular resistance to blood flow is lower in arterioles than in arteries; mean pressure is about 85 mm Hg
c. Capillaries join arterioles and venules
   (1) Their walls, which consists of a single layer of endothelial cells, allow blood and tissue cells to exchange substances
   (2) Vascular resistance to blood flow is very low in capillaries; mean pressure is about 35 mm Hg
d. Venules carry blood from the capillaries to the veins
   (1) Venules have thinner walls than arterioles, but are composed of these same three coats
   (2) Blood pressure is only about 15 mm Hg when blood begins to return to the heart
e. Veins are the largest vessels; they carry blood from the venules to the heart
   (1) Veins have the same three coats as arteries, but the coats are not as thick
   (2) Many veins also have valves, which prevent backflow of blood away from the heart
   (3) Mean pressure in the veins is less than 15 mm Hg

C. Heart

1. The heart is partitioned into two upper chambers with thin muscular walls, the *atria,* and two thicker-walled lower chambers, the *ventricles*
2. *Valves* separate the cardiac chambers from the bases of the aorta and pulmonary artery; they permit blood to flow in one direction only —away from the heart
3. Each side of the heart pumps blood through a different branch of the circulatory system
   a. The right atrium and ventricle pump blood through the pulmonary circulation
   b. The left atrium and ventricle pump blood through the systemic circulation
   c. The muscular walls of the left ventricle are thicker than those of the right ventricle because it pumps blood through the systemic circulation at a much higher pressure than that of the pulmonary circulation
4. The right and left coronary arteries supply blood to the heart muscle; these arteries arise by separate orifices from the part of the aortic wall that is attached to the cusps of the aortic semilunar valves
5. The *cardiac conduction system* sends impulses through the heart muscle to cause synchronized contractions of the atria and ventricles, which pump blood throughout the body

II. Cardiac Cycle

A. General information

1. Impulses generated by the conduction system cause synchronized contractions of the atria and ventricles
2. Each cardiac contraction ***(systole)*** is followed by a period of relaxation ***(diastole);*** the atria and ventricles dilate during their respective diastoles
   a. The atria and ventricles contract and relax in sequence
      (1) *Atrial systole* refers to atrial muscle contraction; atrial diastole refers to relaxation
      (2) *Ventricular systole* refers to ventricular muscle contraction; ventricular diastole refers to relaxation
   b. When used without reference to a specific cardiac chamber, the terms *systole* and *diastole* refer to the ventricular systole and diastole
   c. Events that occur during a single systole and diastole of the atria and ventricles make up the ***cardiac cycle***
3. Actions during the cardiac cycle cause certain characteristic heart sounds and a palpable pulse

B. Systole and diastole

1. When the atria are contracting, the ventricles are relaxed
   a. The *atrioventricular (AV) valves* (valves between the atrial and ventricular chambers) are open during atrial systole
   b. The atria eject blood into the ventricles through the open AV valves
2. When the atria relax, the ventricles contract
   a. Ventricular contraction exerts pressure on the blood in the ventricles, increasing the intraventricular pressure
   b. This rising pressure forces the AV valves to close, preventing blood from flowing backward into the atria
   c. For a short time after the AV valves close, the intraventricular pressure is not high enough to force the *semilunar valves* (valves between the heart and the aorta and pulmonary artery) open; the ventricles are completely closed chambers
   d. Continued ventricular contraction increases the intraventricular pressure until it exceeds the pressure of blood in the aorta and pulmonary artery
   e. This increased intraventricular pressure forces open the semilunar valves and ejects the blood from the ventricles
3. While the ventricles are contracting, venous blood from the systemic and pulmonary circulations is flowing into the relaxed atria
   a. This blood is held in the atria because the AV valves are closed during ventricular systole
   b. As blood fills the atria, the intraatrial pressure rises slightly
4. Pressure in the aorta and pulmonary artery peaks during ventricular systole as the ventricles eject blood, which distends these vessels
5. After expelling this blood, the ventricles begin to relax
   a. The blood ejected from the ventricles into the aorta and pulmonary artery loses its forward momentum caused by the force of ventricular contraction
      (1) Loss of momentum causes the blood to flow back toward the ventricles

(2) Blood fills the cup-shaped cusps of the semilunar valves, forcing the valves shut and preventing blood from refluxing into the ventricles; this maintains high pressure in the aorta and pulmonary artery

(3) The stretched walls of the aorta and pulmonary artery return to their former dimensions by the end of ventricular diastole; this elastic recoil compresses the blood and maintains pressure on it, causing it to continue moving through the vessels in the intervals between ventricular contractions

b. Ventricular pressure falls during ventricular diastole; eventually, it falls below the intraatrial pressure

(1) Then the AV valves open, allowing the accumulated blood in the atria to flow into the ventricles

(2) The same type of pressure changes occur in the right ventricle and pulmonary artery as those in the left ventricle and aorta; however, the pressures are much lower

6. The events of the cardiac cycle are repeated with each heartbeat

C. Normal heart sounds

1. Each cardiac cycle, or heart beat, is associated with characteristic heart sounds ("lub-dub" followed by a pause)
2. These sounds can be heard by placing the ear or a stethoscope against the chest
   a. The "lub" results from vibrations caused by AV valve closure during ventricular systole
   b. The "dub" results from vibrations caused by semilunar valve closure during ventricular diastole
   c. The pause is the interval between successive cardiac contractions

D. Pulse

1. If the heart is contracting normally, pulse assessment is a convenient way to measure the heart rate; the *pulse rate* typically is assessed in a large artery
2. The pulse is produced by a shock wave from ventriculer ejection of blood into the aorta; this shock wave is transmitted through the walls of the large arteries, somewhat like the vibrations transmitted along a metal pipe struck by a hammer
3. The pulse rate is assessed by palpating an artery that lies near the body surface over a bone or other firm tissue and counting the number of pulsations felt in 1 minute

III. Cardiac Electrical Conduction

A. General information

1. Cardiac muscle fibers are polarized; positive charges outside the fibers are balanced by negative charges inside
2. The electrical impulse that causes heart contraction (systole) spreads as a wave through the cardiac muscle; it causes activated areas of the fibers to accumulate negative charges on the outside and positive charges on the inside (depolarization)
3. During diastole, the cell membranes become repolarized as positive charges on the cell surfaces are restored and the cell interiors again accumulate negative charges

B. Conduction system

1. The heart contains a specialized system of nodal tissue for generating and conducting impulses that cause rhythmic contractions
2. The conduction system consists of nodal tissue that contains few myofibrils
3. The cardiac conduction system has four main components: the sinoatrial (SA) node and internodal tracts, atrioventricular (AV) node, the bundle of His, and the Purkinje fibers
   a. The *SA node* is located in the posterior wall of the right atrium near the opening of the superior vena cava
      (1) The SA node is the pacemaker of the heart
      (2) Rhythmic impulses originate here and are conducted to the AV node by small bundles of fibers called internodal tracts
   b. The *AV node* is located in the lower right interatrial septum, near the tricuspid valve (one of the AV valves)
      (1) The impulse slows in the AV node
      (2) Impulse slowing allows the atria to contract and the ventricles to fill with blood
   c. The *bundle of His* originates in the AV node and divides into the right and left bundle branches
      (1) These branches extend down the right and left sides of the interventricular septum
      (2) The impulse from the AV node continues through the bundle of His to the right and left bundle branches
   d. The *Purkinje fibers* connect the right and left bundle branches to the papillary muscles and the lateral walls of the ventricles
      (1) The impulse moves through the Purkinje fibers, eventually reaching the ventricular muscles
      (2) Ventricular muscle stimulation begins in the intraventricular septum and moves downward, causing ventricular depolarization and contraction (see *Cardiac Conduction Route,* page 66, for an illustration)

C. Conduction tracing

1. The electrocardiogram (ECG) traces serial changes in electrical conduction associated with depolarization and repolarization of cardiac muscle fibers
2. These changes are recorded as a series of positive and negative deflections, commonly called waveforms, which produce characteristic patterns (see *Characteristic PQRST Patterns,* page 67, for an illustration)

IV. Blood Pressure

A. General information

1. The pressure of blood in the pulmonary and systemic arteries varies with the phase of the cardiac cycle
2. Pressure is highest when the blood is ejected during systole (systolic pressure) and lowest during diastole immediately before the next cardiac contraction (diastolic pressure)
3. *Blood pressure* refers to the pressure of blood in the systemic circulation, which is about six times higher than the pressure of blood in the pulmonary circulation

Cardiac Conduction Route

In the cardiac conduction system, the impulse begins in the sinoatrial node, travels through the heart chambers, and reaches the ventricular muscle. The illustration below traces the conduction route through these cardiac structures.

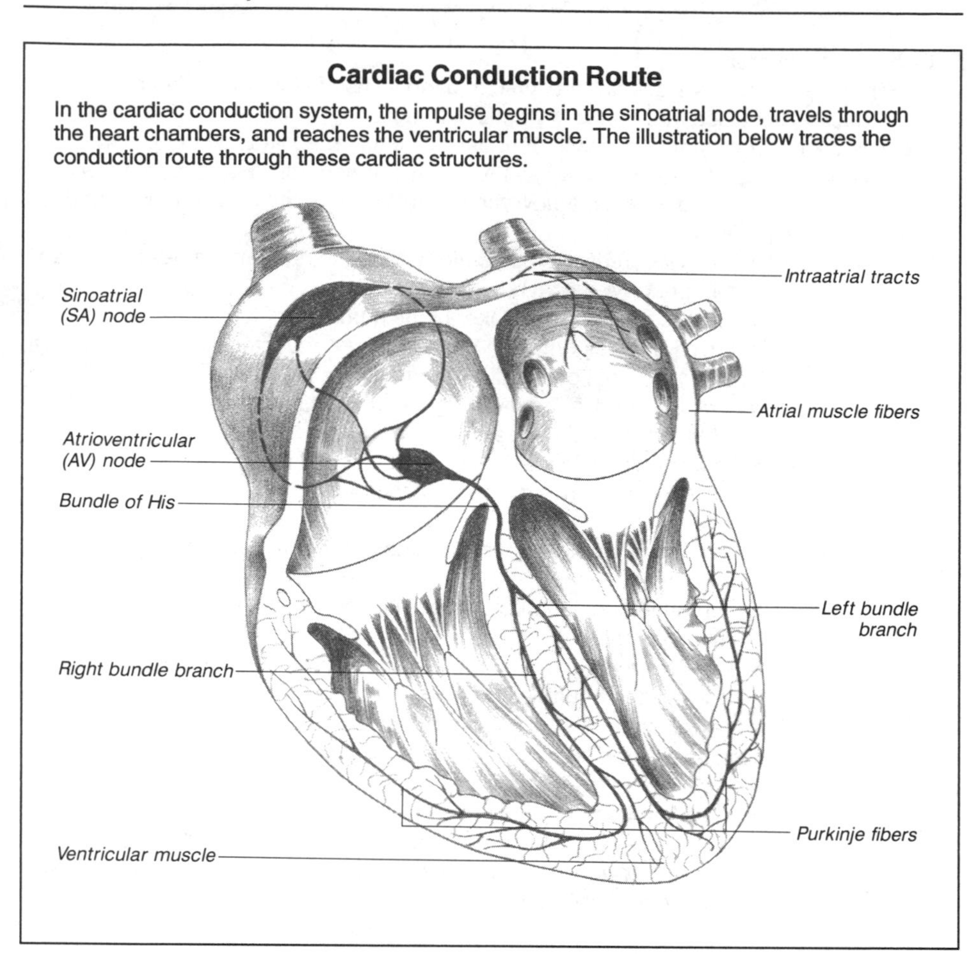

4. A sphygmomanometer and stethoscope can be used to measure systolic and diastolic blood pressure
5. Blood pressure varies rhythmically with the heart beat
    a. The pressure is the highest during ventricular systole when blood is ejected from the ventricles; this represents systolic blood pressure
    b. The pressure is lowest during ventricular diastole when blood is flowing through the arteries into the capillaries; this represents diastolic blood pressure
6. Systolic blood pressure depends primarily on ***cardiac output*** (the force and volume of blood ejected from the ventricles during systole); diastolic pressure depends on ***peripheral resistance*** (degree of impedance to blood flow, which may be increased by vasoconstriction)
7. Blood pressure also is influenced to a lesser degree by blood volume, blood viscosity, and arterial elasticity
8. Systolic pressure is a measure primarily of the force of ventricular contraction; diastolic pressure is a measure of the peripheral resistance caused by arteriolar vasoconstriction

Characteristic PQRST Patterns

Electrocardiogram (ECG) wave patterns are identified by letters; each waveform corresponds to specific electrical events in the cardiac cycle.

- The P wave reflects the initial wave of depolarization associated with atrial systole
- The Q, R, and S waves (collectively called the QRS complex) reflect impulse transmission through the right and left bundles into the terminal branches, leading to ventricular systole
- The T wave reflects ventricular repolarization during diastole
- The P-R interval (time from the beginning of the P wave to the beginning of the QRS complex) represents the time needed for an impulse to pass from the atria to the ventricles through the bundle of His
- The ST segment (time from the end of the S wave to the beginning of the T wave) represents the time between the end of the spread of the impulse through the ventricle and repolarization of the ventricle

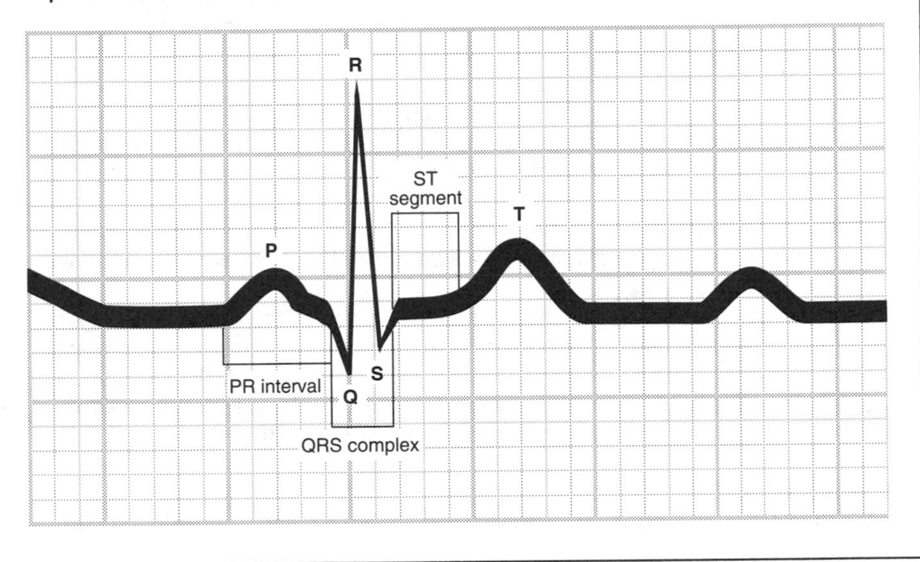

9. Blood pressure does not drop precipitously during diastole; it declines slowly as blood is released gradually into the capillaries by the arterioles, helping to maintain a fairly constant blood pressure
10. A relatively constant blood pressure is necessary to maintain blood flow to tissues and prevent organ damage

B. Cardiac output

1. *Cardiac output* refers to the amount of blood ejected per minute from a ventricle
2. This amount equals the ***stroke volume*** (volume of blood ejected from a single ventricle at each contraction) multiplied by the heart rate in beats per minute (bpm)
3. The basal (resting) stroke volume is about 70 ml; the basal heart rate is about 72 bpm
   a. Therefore, cardiac output is about 5 liters per minute (70 x 72 = 5,040 ml)

b. This amount also equals an individual's total blood volume (approximately 5,000 ml)

4. The ***end-diastolic volume*** (total blood volume in each ventricle before ventricular systole) is about 120 ml
5. The ventricles are not emptied completely of blood during systole
   a. Each ventricle ejects only about 70 ml of blood
   b. The ***end-systolic volume*** (unejected blood remaining in each ventricle) is about 50 ml (120 – 70 = 50 ml)
   c. The ***ejection fraction*** is the amount of blood ejected during each ventricular contraction in relation to the end-diastolic volume
   d. In this example, 70 ml (stroke volume) is divided by 120 ml (end-diastolic volume), which equals an ejection fraction of 0.58 or 58%
6. The heart can increase its rate and stroke volume as needed to increase cardiac output above the basal level
   a. This increase can be fourfold, if necessary
   b. Trained athletes can increase their cardiac output even more
   c. If cardiac output increases, blood pressure tends to rise because blood is delivered to the arteries more rapidly than it leaves through the arterioles

C. Peripheral resistance

1. The degree of peripheral resistance is determined by the degree of arteriolar vasoconstriction
2. Arterioles restrict blood flow from the arteries into the capillaries
   a. If arteriolar constriction and peripheral resistance increase, blood pressure rises because blood flow from the arteries into the capillaries is impeded
   b. If the arterioles relax and peripheral resistance decreases, blood flows more rapidly into the capillaries and arterial pressure drops

D. Blood volume

1. Blood pressure tends to vary directly with the total blood volume in the cardiovascular system
2. Blood pressure tends to rise if blood volume increases (which occurs in some diseases) and tends to drop if blood volume falls below normal (which occurs in severe hemorrhage)

E. Blood viscosity

1. The greater the viscosity of a fluid, the more slowly it flows through a narrow orifice; the reverse also is true
2. In some diseases, blood viscosity rises as a result of an increase in proteins in the blood or in the number of erythrocytes
   a. These conditions restrict blood flow through arterioles into capillaries
   b. As a result, blood pressure tends to rise

F. Arterial elasticity

1. Blood pressure changes if the elasticity of the large arteries changes
   a. Normally, large artery distention absorbs some of the force of the blood ejected during systole
   b. Distention prevents excessive elevation of systolic pressure during each ventricular contraction
   c. The arteries recoil during diastole and propel the blood forward

2. In some diseases, the aorta and large arteries gradually lose their elasticity and become more rigid
   a. Then they cannot distend and absorb the impact of the ejected blood
   b. As a result, systolic pressure tends to rise

V. Regulation of Cardiac Output and Blood Pressure

A. General information

1. Adequate cardiac output and stable blood pressure are essential for proper performance of major body organs
2. Cardiac output normally varies in response to the body's requirements; blood pressure normally remains within a specific range
3. Several mechanisms control cardiac output and blood pressure
   a. The heart exerts control over its stroke volume, as described by ***Starling's law***
   b. The autonomic nervous system controls nerve impulse discharge based on information relayed to the brain from baroreceptors and chemoreceptors in major arteries
   c. The kidneys secrete hormones that affect the heart

B. Starling's law

1. According to Starling's law, the amount of stretching of cardiac muscle fibers helps regulate stroke volume and maintain equal output from both ventricles
2. Muscle stretching commonly occurs in two ways
   a. If venous return to the ventricle increases, the ventricle becomes overdistended and its muscles become stretched
   b. If the diastolic pressure rises, the heart must eject blood against a higher peripheral resistance; when this occurs, the ventricle tends to empty less completely and becomes overdistended, stretching the cardiac muscles
3. Stretching of the fibers causes the ventricle to contract more forcefully to expel the additional blood
4. The more forceful ventricular contractions maintain normal cardiac output and supply adequate blood to the tissues despite increased peripheral resistance

C. Regulation by baroreceptors

1. The autonomic nervous system (ANS) controls the heart and blood vessels
2. A diffuse network of interconnecting neurons in the medulla regulate the discharge of autonomic nerve impulses, primarily in response to impulses from *baroreceptors* (specialized pressure-sensitive receptors in the carotid artery and walls of the aortic arch)
3. This medullary control center exerts opposing effects
   a. Its cardioacceleratory portion controls nerve impulse discharge from the sympathetic division of the autonomic nervous system, which speeds the heart rate
   b. Its cardioinhibitory portion controls nerve impulse discharge from the parasympathetic division, which slows the heart rate
   c. The net effect of the ANS on the heart and blood vessels reflects the combined response of the two portions of the medullary control center

4. In response to input from baroreceptors, the medullary centers also send regulatory impulses to the heart and blood vessels, which maintain a stable cardiac output and blood pressure
   a. If the intravascular pressure rises, baroreceptors send impulses to the medullary center
      (1) These impulses initiate discharge of parasympathetic impulses in the vagus nerve to slow the heart rate and reduce the force of ventricular contraction, which reduces cardiac output, and to reduce arteriolar vasoconstriction, which reduces intravascular pressure
      (2) Impulses simultaneously inhibit sympathetic output from the cardiaoacceleratory portion of the center
   b. If the intravascular pressure falls, baroreceptors relay fewer impulses to the medullary center
      (1) The medullary center responds by discharging sympathetic nerve impulses to increase the heart rate and force of ventricular contraction, which increases cardiac output, and to constrict the arterioles, which raises intravascular pressure
      (2) Impulses simultaneously inhibit parasympathetic output from the cardioinhibitory portion of the center

D. Regulation by chemoreceptors

1. Chemoreceptors (chemical-sensitive receptors in the aortic arch and carotid sinus) respond to decreases in blood oxygen concentration and pH; they also respond to increases in blood carbon dioxide concentration
2. The medullary center initiates impulses via the ANS that increase heart rate and blood pressure
3. Chemoreceptors do not play a major role in heart rate and blood pressure regulation under normal physiologic conditions; however, they respond to extreme PO_2 and blood pH decreases and extreme PCO_2 increases
4. They transmit impulses to the medullary control center via the same pathways that transmit impulses from baroreceptors

E. Hormonal regulation

1. In the kidneys, the hormones renin, angiotensin, and aldosterone help regulate blood pressure; the ***renin-angiotensin-aldosterone mechanism*** responds more slowly to blood pressure changes than the reflex mechanisms
2. A sustained fall in blood pressure causes the kidneys to release renin, which is converted to angiotensin in the circulation
3. Angiotensin raises blood pressure by causing arteriolar constriction; it also stimulates the adrenal gland to release the steroid hormone aldosterone
4. Aldosterone promotes sodium and water retention by the kidneys; this retention increases blood volume and leads to a corresponding increase in blood pressure

VI. Capillary and Interstitial Tissue Circulation

A. General information

1. Capillary blood and cells exchange such substances as oxygen, nutrients, and cellular waste products
2. These substances must pass in solution through the loose connective tissue (interstitial space) in which the cells are dispersed
3. The interstitial space is filled with a semisolid matrix containing connective tissue fibers and a small amount of interstitial fluid that has filtered through the capillaries
4. At the arterial end of the capillary, the ***hydrostatic pressure*** (which pushes fluid out) is higher than the osmotic pressure (which pulls fluid back), causing interstitial fluid to filter through the endothelium of the capillary into the interstitial space
5. At the venous end of the capillary, the hydrostatic pressure is lower than the osmotic pressure, causing fluid to diffuse back into the capillary
6. *Edema* (excess fluid in the interstitial space) may result from any disturbance in fluid transfer between the capillaries and interstitial tissue

B. Fluid movement

1. The cyclic flow of fluid from the interstitial space to the capillaries and back into the interstitial space depends on capillary hydrostatic pressure, capillary permeability, osmotic pressure, and open lymphatic channels
2. Blood in the capillaries exerts capillary hydrostatic pressure
   a. This pressure tends to force fluid out from the blood through the capillary endothelium and into the interstitial fluid
   b. This pressure is much lower than that in the larger arteries because arteriolar constriction restricts blood flow into the capillaries, functioning like a pressure reduction valve
   c. As a result, the mean arterial pressure in larger arteries falls from about 85 mm Hg to about 35 mm Hg at the arterial end of the capillary and drops to a low of 15 mm Hg at the venous end of the capillary
3. Capillary permeability determines the ease with which fluid can pass through the capillary endothelium
4. The osmotic pressure exerted by the blood proteins (called the colloid osmotic pressure) tends to attract interstitial fluid back into the capillaries
5. Open lymphatic channels collect some fluid forced out into the interstitial tissues and return it to the circulation; this occurs because the pressure in the lymphatic channels is lower than that in the interstitial tissues
6. Edema may result from any disturbance in the factors that regulate fluid transfer in the capillary–interstitial tissue–capillary cycle; causes of such disturbance include inflammation, decreased colloid osmotic pressure, increased capillary hydrostatic pressure, and lymphatic channel obstruction

Study Activities

1. Identify the components of the cardiovascular system and explain their basic functions.
2. Trace the blood flow through the heart during the cardiac cycle.
3. Describe the parts of the cardiac conduction system and explain how they work together to stimulate ventricular contraction.
4. Identify cardiac conduction events in relation to an ECG.
5. Differentiate between systolic and diastolic blood pressures.
6. Discuss the major factors that affect blood pressure.
7. Explain the mechanisms that regulate cardiac output and blood pressure.
8. Describe how the body maintains capillary and interstitial tissue circulation.

7

Blood

Objectives

After studying this chapter, the reader should be able to:

- Describe the basic functions of the formed elements and plasma.
- Trace the life cycle of an erythrocyte.
- Discuss erythrocyte maturation, including the hemoglobin synthesis.
- Discuss the structure and function of iron.
- Describe the structure and function of hemoglobin.
- Compare the different types of hemoglobin, noting the globin chain composition and normal quantity of each.
- Explain the process of blood coagulation, including the factors involved.
- Explain the process of fibrinolysis, including the factors involved.
- Describe the role of genetics, antigens, and antibodies in the ABO and Rh blood group systems.

I. Blood Composition and Functions

A. General information

1. Blood consists of *formed elements (erythrocytes, leukocytes,* and *thrombocytes* or *platelets)* suspended in a viscous fluid (blood *plasma*)
2. Blood performs several functions
   a. It transports oxygen and nutrients to cells
   b. It removes carbon dioxide and other cellular wastes, promoting *homeostasis* (natural balance in the internal environment of the body)
   c. It carries the formed elements of blood and products of cell metabolism (including hormones and antibodies) throughout the body
   d. It helps protect the body from invasion by foreign substances, such as viruses and antibodies
   e. It prevents excessive blood loss through hemostasis
3. The body of an average adult contains about 5 liters (5 quarts) of blood

B. Plasma

1. Plasma contains about 7% proteins and about 93% water with dissolved minerals, nutrients (such as glucose and amino acids), and cellular wastes
2. Plasma proteins consist of *albumin, globulins,* and *fibrinogen*
   a. Albumin comprises about half the total plasma proteins
      (1) It is essential in maintaining the colloid osmotic pressure of the blood

(2) This pressure plays a major role in regulating fluid flow between the capillaries and interstitial tissues (see Chapter 6, Cardiovascular System, for more information)

b. Globulins are divided into alpha, beta, and gamma globulins based on their physical and chemical characteristics; each type of globulin has a specific function

(1) Some alpha and beta globulins are concerned with blood coagulation; others transport enzymes, hormones, vitamins, and other substances

(2) Gamma globulins act as antibodies, defending the body against infection

c. Fibrinogen plays a major role in blood coagulation; it is converted to fibrin when the blood coagulates

3. *Serum* is the fluid expressed from a clot

a. Like plasma, serum contains no formed elements

b. Unlike plasma, serum lacks fibrinogen and other proteins that are depleted by blood coagulation

C. Formed elements

1. Formed elements of blood consist of erythrocytes (red blood cells), leukocytes (white blood cells), and thrombocytes (platelets)
2. Each type of cell has a characteristic structure and staining reaction when examined microscopically
3. All formed elements develop from a common precursor cell, the stem cell, which is found in the red bone marrow of certain bones
   a. Stem cells transform into immature versions of each formed element, which then mature
   b. This process, called ***hematopoiesis,*** produces the various types of blood cells (see *Hematopoeisis* for an illustration)
4. The quantity of each type of blood cell in circulation is maintained within narrow limits, which vary with the individual's age and physical demands
   a. Blood normally contains 4.5 to 5.5 million erythrocytes per mm^3
   b. Blood normally contains 5,000 to 10,000 leukocytes per mm^3
   c. Blood normally contains 150,000 to 350,000 platelets per mm^3

II. Erythrocytes

A. General information

1. Erythrocytes are the most numerous formed element in the blood, averaging about 5 million per mm^3
   a. These cells are flexible, biconcave disks that measure about 7 microns in diameter (1 mm = 1,000 microns)
   b. They lack nuclei, which are lost as the cells mature
2. Erythrocyte maturation requires *hemoglobin* formation
3. A protein vital to erythrocyte function, hemoglobin is composed of *heme* and *globin*
   a. Heme is an iron-containing porphyrin compound
   b. Globin is a protein composed of amino acids
4. The mature erythrocyte consists of a hemoglobin solution enclosed in a protein framework and surrounded by a lipoprotein cell membrane

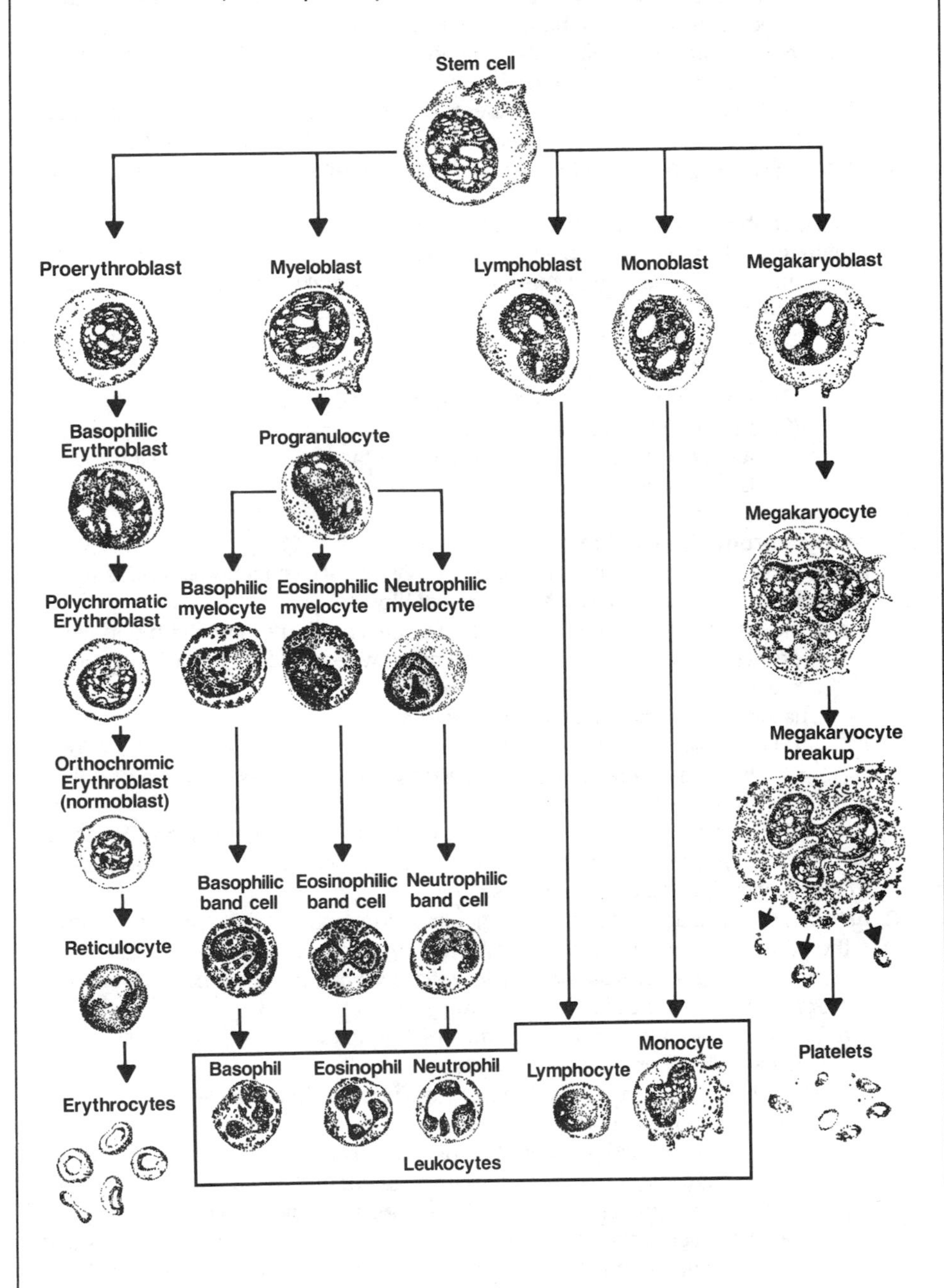
Hematopoeisis
From one stem cell, hematopoiesis produces all seven types of blood cells as shown below.
Stem cell
Proerythroblast
Myeloblast
Lymphoblast
Monoblast
Megakaryoblast
Basophilic Erythroblast
Progranulocyte
Megakaryocyte
Basophilic myelocyte
Eosinophilic myelocyte
Neutrophilic myelocyte
Polychromatic Erythroblast
Megakaryocyte breakup
Orthochromic Erythroblast (normoblast)
Basophilic band cell
Eosinophilic band cell
Neutrophilic band cell
Reticulocyte
Monocyte
Platelets
Basophil
Eosinophil
Neutrophil
Lymphocyte
Erythrocytes
Leukocytes

a. The mature erythrocyte circulates through the cardiovascular system, carrying oxygen to tissues and removing carbon dioxide
b. It contains enzymes that allow it to perform essential metabolic functions
5. An erythrocyte lacks mitochondria and cannot obtain energy from mitochondrial enzyme systems like other cells; instead, it derives energy from glucose breakdown via other enzyme pathways
6. The mature erythrocyte cannot synthesize proteins because it lacks a nucleus; therefore, it cannot make new enzymes to replace those that wear out
7. Erythrocyte aging results in part from progressive deterioration of its enzyme systems
   a. After about 4 months, cells no longer can function
   b. In the spleen, the mononuclear phagocyte (reticuloendothelial) system removes aged cells from the circulation and breaks them down into components that are recycled or discarded
      (1) Globin chains are broken down and their component amino acids are used to make other proteins
      (2) Iron is extracted and recycled to make new hemoglobin
      (3) In the heme molecule, the porphyrin ring cannot be salvaged; instead, it is broken down and transported to the liver, where it is excreted as bile pigment (bilirubin)

B. Erythrocyte production and maturation

1. Erythrocyte production occurs in the bone marrow of certain bones and is regulated by the oxygen content of the arterial blood
   a. If the number of circulating erythrocytes falls below normal, the blood delivers less oxygen to the tissues; the marrow responds to this oxygen decrease by increasing erythrocyte output
   b. The kidneys mediate this marrow response
      (1) Specialized kidney cells respond to low oxygen tension by liberating the enzyme *renal erythropoietic factor,* which acts on blood protein to form *erythropoietin*
      (2) Erythropoietin stimulates the stem cells in the bone marrow to produce erythroblasts, which tend to form in groups called *erythroblast islands*
2. Erythroblasts develop over several stages
3. As each erythroblast matures, hemoglobin accumulates in the cytoplasm, and the nuclear chromatin condenses; normal erythrocyte maturation is called *normoblastic* ***erythropoiesis*** to differentiate it from abnormal maturation, or *megablastic eythropoeisis*
   a. When the cell has synthesized most of its hemoglobin, it extrudes its nucleus and becomes a *reticulocyte*
   b. Then the reticulocyte enters one of the sinusoids in the bone marrow and is carried into the circulatory system
      (1) In a microscopic examination of stained blood, the reticulocyte appears slightly larger than a mature erythrocyte and has a faint blue color because it contains less red-staining hemoglobin than an erythrocyte
      (2) The reticulocyte also can be distinguished by using special stains that precipitate and clump the mitochondria and other organelles within the cytoplasm and cause them to appear as a meshwork (reticulum) of irregular blue-staining strands and granules in the cytoplasm; the name *reticulocyte* is derived from this meshwork

4. Final cell maturation occurs in the circulation
   a. After losing its nucleus, the cell retains its organelles and continues to synthesize additional hemoglobin for the next 24 hours
   b. The reticulocyte matures into an erythrocyte over the next 24 hours; the cell gradually decreases in size and loses its blue tinge as more hemoglobin is synthesized
5. The proportion of reticulocytes serves as a useful index of erythrocyte production
   a. Because erythrocytes survive in the circulation for about 100 days (actually 120 days, but considered as 100 days to simplify calculations), approximately 1% of the body's erythrocytes must be produced each day to replace those that wear out and are eliminated from the circulation
   b. The normal reticulocyte percentage in the circulation, therefore, is approximately 1%
   c. Erythrocyte production increases to compensate for increased blood loss (such as after blood donation) or reduced erythrocyte survival (as in some types of anemia); the percentage of circulating reticulocytes also increases correspondingly

C. Iron structure and formation

1. Iron forms an essential part of hemoglobin and also appears in *myoglobin,* a similar compound in muscle
2. The body contains about four grams of iron; hemoglobin contains about 75% of this amount
3. Additional iron is stored with the protein *apoferritin* in the bone marrow, liver, and spleen, forming two different iron-protein complexes
   a. Ferritin is about 20% iron
   b. Hemosiderin is about 25% iron
4. Small amounts of iron circulate in the plasma with the iron-binding transport protein *transferrin;* this protein transports iron from the intestinal tract (where iron is absorbed) and from mononuclear phagocytes (where iron is recovered from erythrocyte breakdown) to the bone marrow, liver, and spleen (where extra iron is stored as ferritin and hemosiderin)
5. Only small amounts of iron are absorbed and excreted; intestinal mucosal cells closely control the body's iron absorption to ensure an adequate—but not excessive—iron supply
   a. Dietary iron absorption occurs principally in the duodenum; it is regulated by the iron content of the intestinal mucosal cells
   b. When iron stores are abundant, the mucosal cells contain ferritin, which inhibits their uptake of additional iron
   c. When iron stores are low, the mucosal cells absorb more iron, which enters the plasma and is carried by transferrin to the bone marrow and storage sites until it is needed for hemoglobin synthesis
6. The usual diet contains about 10 to 20 mg of iron
   a. Men absorb about 1 mg per day
   b. Women and children absorb only slightly more
      (1) Women require more iron to compensate for blood loss during menstruation and for fetal needs during pregnancy
      (2) Children require extra iron to synthesize hemoglobin during periods of growth when blood volume increases

7. Pregnancy, childhood growth, or chronic or excessive blood loss may deplete iron stores; because dietary iron intake may not be sufficient to compensate for this depletion, iron-deficiency anemia may result
8. Normally, iron from erythrocyte breakdown is recycled to make new hemoglobin
9. One ml of blood contains approximately 0.5 mg of iron

D. Hemoglobin structure and function

1. Hemoglobin is the oxygen-carrying protein formed in the cytoplasm of developing erythrocytes
2. It is composed of four monomers fitted together to form a tetramer, like the sections of an apple cut into quarters
3. Each monomer is a complex composed of the iron-containing compound heme and the protein globin
    a. Heme is constructed from four nitrogen-containing ring compounds called pyrrol rings
    b. These rings are joined to form a more complex ring structure called a porphyrin ring
4. An atom of iron, with six binding sites, occupies the central position of the porphyrin ring
    a. Nitrogen atoms bind to four of the sites on the porphyrin ring
    b. The amino acid histidine binds the globin chain to the iron at the fifth site
    c. The sixth binding site is available for reversible combination with oxygen
5. Globin forms the largest part of the hemoglobin monomer
    a. It is composed of an amino acid chain joined together to form a coiled polypeptide chain
    b. The heme is tucked into one of the bends in the globin coil and held there by the bond between the iron atom and histidine in the globin chain
6. Heme and globin are synthesized in different locations in the cell
    a. Transferrin brings iron to the erythroblast
    b. Once inside the mitochondria, iron is incorporated into the porphyrin ring to form heme (nearby reticuloendothelial cells remove excess iron that is not used in heme synthesis)
    c. At the same time, ribosomes in the cytoplasm produce globin chains
    d. The globin chains and heme join to form a hemoglobin monomer
    e. The four monomers aggregate to form the complete hemoglobin tetramer

E. Globin chains

1. The five types of globin chains —alpha, beta, gamma, delta, and epsilon —have different amino acid compositions
2. Globin chains are produced at different times and in different proportions, beginning in the embryonic period and extending into adult life
3. In most hemoglobin tetramers, two of the subunits contain one type of globin chain and the other two subunits have a different globin chain
    a. Each of the four chains is combined with heme to form a hemaglobin monomer
    b. Four monomers aggregate to form a tetramer
4. Adults produce two types of hemoglobin
    a. In adults, about 98% of the hemoglobin consists of tetramers in which two subunits contain alpha chains and the other two contain beta chains; this is *hemoglobin A* or adult hemoglobin

b. The other 2% of hemoglobin consists of tetramers with alpha and delta chains; this is *hemoglobin A_2*

5. The embryo and fetus produce different hemoglobins
   a. The embryo initially produces hemoglobin that contains only epsilon chains
   b. The fetus soon replaces this tetramer with production of a tetramer that contains alpha and gamma chains; this is *hemoglobin F* or fetal hemoglobin, which is the predominant hemoglobin in the fetus
   c. Late in the fetal period, the beta chains of adult hemoglobin are produced in small quantities
6. Fetal hemoglobin can take up and release oxygen at lower oxygen tensions than hemoglobin A; this is advantageous because the oxygen tension in the fetal blood is relatively low
7. After birth, hemoglobin production is characterized by declining gamma chain synthesis and increasing beta chain synthesis
   a. Hemoglobin F in the erythrocytes gradually declines
   b. Hemoglobin A rises correspondingly
8. Genes control the globin chain structure by directing the order in which amino acids are incorporated into the chains; mutation of these genes can alter the amino acid sequence, causing hemoglobin abnormalities, such as sickle cell anemia

III. Leukocytes

A. General information

1. Five different types of leukocytes exist: *neutrophils, eosinophils, basophils, lymphocytes,* and *monocytes*
2. Each type has specific functions
3. Leukocytes are less numerous than erythrocytes; the average leukocyte count is about 7,000/mm^3, but may vary widely based on individuals' needs
4. Leukocytes undergo maturation in the bone marrow
   a. Leukocyte maturation is characterized by condensation of nuclear chromatins and an increase in cytoplasm
   b. Several types of leukocytes develop different kinds of cytoplasmic granules and multilobed nuclei
5. Most leukocytes have a short life span
   a. They circulate about 4 to 8 hours after they are released from the bone marrow; then they leave the circulation and enter the tissues
   b. They live another 4 to 5 days in various tissues
   c. The total life span may decrease to a few hours in an individual with a severe infection

B. Neutrophils

1. Neutrophils (polymorphonuclear leukocytes) have a segmented nucleus and a cytoplasm containing fine granules
2. They are the most numerous leukocytes, comprising about 65% of all leukocytes
3. These cells are phagocytic and can destroy bacteria, viruses, and other foreign substances
4. Their numbers increase in response to infections

C. Eosinophils

1. Eosinophils contain large, bright red-staining (eosinophilic) granules
2. They account for about 2% of all leukocytes
3. Their major function is phagocytosis of antigen-antibody complexes; these complexes form in allergic reactions when antibodies combine with the antigens that triggered their release
4. The number of eosinophils increases in many allergic conditions and in response to some parasitic infections

D. Basophils

1. Basophils contain dark purple (basophilic) granules
2. They account for less than 1% of all leukocytes
3. Basophils play an essential role in some allergic reactions
4. In a sensitized individual, specific antigens, such as penicillin and bee venom, cause basophils to rupture
   a. Basophil rupture releases large quantities of histamine, bradykinin, heparin, and some lysosomal enzymes
   b. These substances produce the tissue and vascular reactions that cause allergy signs and symptoms

E. Lymphocytes

1. Lymphocytes have a deep-staining nucleus that contains dense chromatin and a pale blue-staining cytoplasm
2. Lymphocytes normally comprise about 30% of leukocytes
3. Lymphocytes play a major role in the body's immune response (see Chapter 8, Lymphatic System and Immunity, for more information on lymphocyte function)
4. Lymphocytes can be divided into two groups based on their survival length
   a. One group has a short survival comparable to other leukocytes
   b. The other lymphocyte group survives for several years; the cells in this group are known as *memory cells*
5. Lymphocytes increase in number in response to some infections

F. Monocytes

1. Monocytes are phagocytic cells that form part of the mononuclear phagocyte system
2. They account for about 3% of all leukocytes
3. They leave the blood stream and function as scavenger cells, cleaning up the debris after acute inflammation
4. Monocytes also function as the body's chief defense against some chronic diseases, such as tuberculosis, and some systemic fungal infections
5. They participate along with the lymphocytes in the body's immune response (see Chapter 8, Lymphatic System and Immunity, for more information)

IV. Thrombocytes

A. General information

1. Thrombocytes (platelets) play an essential role in blood clotting; inadequate platelet quantity or function can cause serious bleeding problems

2. Platelets are small structures, about one-third the size of erythrocytes; like erythrocytes, they have no nucleus

B. Platelet life cycle

1. Platelets are formed from *megakaryocytes* in the bone marrow
   a. Megakaryocytes are extremely large cells that develop from stem cells
   b. On the surface of megakaryocytes, platelets form as buds that pinch off to enter the circulation
   c. Blood normally contains 150,000 to 350,000 platelets per mm^3
2. Platelets are motile and can store and release biochemicals and change shape
3. Their lifespan is about 10 to 14 days

V. Hemostasis

A. General information

1. ***Hemostasis*** refers to stoppage of bleeding; the body arrests all bleeding as quickly as possible to prevent life-threatening hemorrhage
2. Blood vessels, platelets, and coagulation factors help blood to clot, which stops the bleeding
3. After the blood vessel has healed and the clot no longer is needed, it must be lysed (dissolved)
4. The body normally maintains a balance between clotting and clot ***fibrinolysis,*** which is controlled largely by substances that inhibit and activate clotting and lysis

B. Role of blood vessels and platelets

1. Blood vessels and platelets function together to prevent bleeding
2. Injury to a blood vessel causes it to constrict, narrowing its diameter and facilitating closure by a blood clot
3. Vessel injury also disrupts the endothelium and exposes the underlying connective tissue, stimulating platelets and coagulation factors to react
4. Platelets aggregate and adhere to the injury site
   a. They release vasoconstrictors, causing vessels to constrict further
   b. These platelets also release platelet phospholipid that initiates ***coagulation***
5. Platelets also play an important role in preventing bleeding from capillaries
   a. Small breaks in capillaries occur frequently, but are sealed promptly by formation of a platelet plug rather than a blood clot
   b. If the number of platelets is inadequate or if platelets do not function properly, pinpoint areas of bleeding (petechiae) develop in the skin and internal organs and may be followed by more serious bleeding

C. Plasma coagulation factors

1. Coagulation factors are designated by name and Roman numeral (see *Coagulation Factors,* page 82, for a list)
2. They circulate as precursor compounds and are activated during coagulation
3. Coagulation is a chain reaction in which each coagulation factor is activated in sequence and in turn activates the next coagulation factor in the chain

Coagulation Factors

Various factors perform vital functions in the coagulation process, as described in the chart below.

FACTOR NUMBER AND NAME	DESCRIPTION AND FUNCTION
I (fibrinogen)	High-molecular-weight protein synthesized in liver; converted to fibrin in phase 3 (a common phase)
II (prothrombin)	Protein synthesized in liver (requires vitamin K); converted to thrombin in phase 2 (a common phase)
III (tissue thromboplastin)	Factor released from damaged tissue; required in phase 1 of the extrinsic system
IV (calcium ions)	Factor required throughout the entire clotting sequence
V (proaccelerin or labile factor)	Protein synthesized in liver; functions in phases 1 and 2 of the intrinsic and extrinsic systems
VII (serum prothrombin conversion accelerator, stable factor, or proconvertin)	Protein synthesized in liver (requires vitamin K); functions in phase 1 of the extrinsic system
VIII (antihemophilic factor or antihemophilic globulin)	Protein synthesized in liver; required in phase 1 of the intrinsic system
IX (plasma thromboplastin component)	Protein synthesized in liver (requires vitamin K); required in phase 1 of the intrinsic system
X (Stuart factor or Stuart-Prower factor)	Protein synthesized in liver (requires vitamin K); required in phase 1 of the intrinsic and extrinsic systems
XI (plasma thromboplastin antecedent)	Protein synthesized in liver; required in phase 1 of the intrinsic system
XII (Hageman factor)	Protein required in phase 1 of the intrinsic system
XIII (fibrin stabilizing factor)	Protein required to stabilize the fibrin strands in phase 3

4. Several coagulation factors are manufactured in the liver and require vitamin K for their synthesis; vitamin K deficiency can inhibit their synthesis and disrupt normal coagulation
5. All phases of coagulation require calcium ions; however, coagulation disturbances do not result from abnormally low blood calcium levels because a calcium level low enough to affect coagulation would be incompatible with life

D. Coagulation factor activation

1. Although the coagulation cascade is a continuous sequence, it can be divided into three phases for discussion purposes (see *Coagulation Cascade,* page 84, for a diagram)
2. Phase 1 involves formation of prothrombin activator
   a. Prothrombin activator is produced by two different mechanisms: the intrinsic system and the extrinsic system
   b. The intrinsic system is so named because all of its components come from the blood

(1) Platelets accumulate at the injury site and release phospholipids, which interact with coagulation factors

(2) The factors are activated in sequence to yield prothrombin activator

c. In the extrinsic system, tissue injury liberates thromboplastin (Factor III), which interacts with coagulation factors to produce Factor X_a

(1) The extrinsic system is so named because prothrombin activator is derived primarily from injured tissue rather than blood

(2) This system requires about half the number of coagulation factors as the intrinsic system; it requires no platelets

3. Phase 2 involves conversion of prothrombin (Factor II) to thrombin by prothrombin activator, which was formed in phase 1
   a. Prothrombin is a protein made in the liver
   b. Prothrombin activator splits prothrombin into fragments
   c. Fragmentation produces thrombin, a protein-digesting enzyme
4. Phase 3 involves conversion of fibrinogen (Factor I) to fibrin by thrombin
   a. Fibrinogen is a high-molecular-weight protein produced by the liver
   b. Thrombin cleaves part of the fibrinogen molecule to form a smaller molecule, called a fibrin monomer
   c. Then fibrin monomers join end-to-end and cross link from side-to-side to form a meshwork of fibrin threads that contain plasma, erythrocytes, leukocytes, and platelets; this meshwork is a clot
   d. Another plasma factor (XIII) strengthens the bonds between the fibrin molecules and increases the strength of the clot

E. Coagulation inhibitors

1. Coagulation factors and mechanisms are counterbalanced by inhibitors and anticoagulants that retard clotting, which prevents excessive intravascular coagulation
2. Antithrombin in plasma inhibits thrombin formation
3. Heparin, which exists in basophils and mast cells, has anticoagulant properties
4. Prostaglandin derivatives inhibit the platelet aggregation and phospholipid release that initiate coagulation

F. Fibrinolytic system

1. This clot-dissolving system is activated simultaneously with the coagulation mechanism; it restricts clotting to a limited area, thereby preventing excessive intravascular coagulation
2. The precursor compound plasminogen (profibrinolysin) is converted to plasmin (fibrinolysin) by various substances
   a. Factor XII_a, which formed during phase 1 of coagulation by the intrinsic system, activates plasminogen conversion
   b. Components from injured tissues activate the extrinsic system as well as plasminogen conversion
   c. Thrombin, which converts fibrinogen to fibrin, also converts plasminogen to plasmin
3. Plasmin is a proteolytic (protein-digesting) enzyme that breaks down the fibrin strands in the clot
4. Plasmin is inactivated by circulating inhibitors called alpha globulins

Coagulation Cascade

Via the intrinsic and extrinsic systems, various factors interact and lead to coagulation.

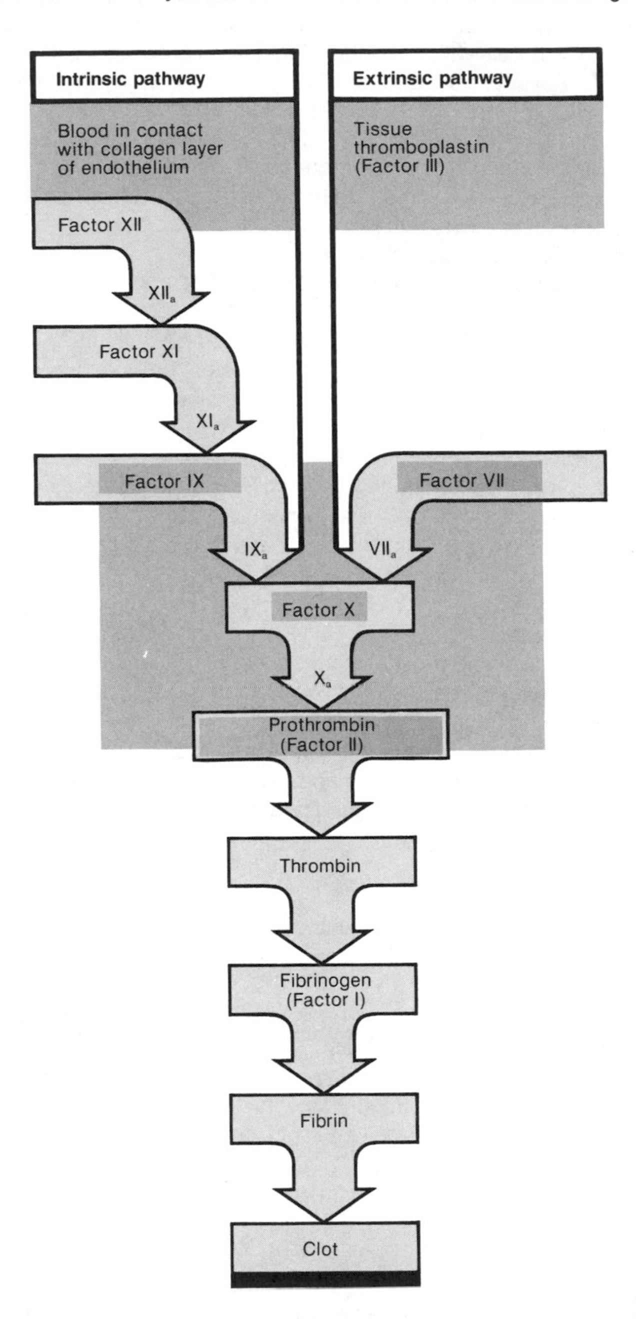

Adapted from Baer, C., and Williams, B. (1991). *Clinical Pharmacology and Nursing,* (2nd ed.). Springhouse, PA: Springhouse Corporation.

a. Alpha globulins combine with plasmin to form inactive complexes; this prevents unwanted clot breakdown by inactivating plasmin trapped in the fibrin clot
b. These inhibitors also inactivate plasmin that diffuses from the coagulation sites; this prevents plasmin from attacking and degrading other clotting factors

VI. Blood Groups

A. General information

1. The surface of erythrocytes contains antigens that are inherited through Mendelian patterns
2. When exposed to certain antibodies, these antigens cause reactions that are the basis for classifying blood groups
3. Many different blood group systems have been described; the ABO and Rh systems are the most important

B. ABO blood group system

1. Erythrocytes are classified primarily on the presence or absence of A and B antigens on their surfaces and secondarily on the presence or absence of anti-A and anti-B antibodies in the serum
2. Individuals normally have antibodies directed against antigens their erythrocytes lack
   a. Group A individuals (41% of the population) have A antigen on the erythrocytes and anti-B antibody in the serum
   b. Group B individuals (10% of the population) have B antigen on the erythrocytes and anti-A antibody in the serum
   c. Group O individuals (45% of the population) have no A or B antigens on the erythrocytes; they have anti-A and anti-B antibodies in the serum
   d. Group AB individuals (4% of the population) have both A and B antigens on the erythrocytes and no anti-A or anti-B antibodies in the serum
3. ABO blood group (blood type) is determined by identifying the ABO antigens on the erythrocytes and the anti-A and anti-B antibodies in the serum
4. Antigens are identified by mixing a sample of the erythrocytes with anti-A typing serum and another sample with anti-B typing serum; this ABO antigen typing is known as *direct typing*
   a. Erythrocyte clumping ***(agglutination)*** in the typing serum determines the antigens
   b. For example, if the cells agglutinate with anti-A typing serum, but not anti-B typing serum, then the individual's erythrocytes contain the A antigen
5. Serum antibodies are determined by mixing one serum sample with a suspension of known group A test cells and another sample with a suspension of known group B test cells; this antibody typing is known as *reverse typing* and is used to confirm direct typing
   a. Test cell agglutination indicates the corresponding antibody in the serum
   b. For example, if the serum sample agglutinates group B test cells but not group A test cells, the serum contains only anti-B antibody
6. ABO antigens are determined genetically

a. The gene locus on each chromosome may be occupied by an A, B, or O gene; these genes are transmitted to offspring by Mendelian inheritance patterns
   (1) An individual is homozygous for ABO genes if the alleles are the same on both chromosomes (AA, BB, or OO)
   (2) An individual is heterozygous if the genes are different (AO, BO, AB)
b. The gene combination on the chromosomes determines the ABO *genotype*
c. The blood type obtained by testing erythrocytes with anti-A and anti-B typing serum determines the ABO *phenotype*
   (1) A phenotype may include more than one genotype
   (2) For example, the phenotype B may represent either genotype BB (homozygous) or BO (heterozygous)
d. Because parents may transmit either allele to their offspring, an infant may not have the same ABO genotype or phenotype as either parent
   (1) For example, two heterozygous type A (AO) parents may produce a homozygous type A (AA), heterozygous type A (AO), or type O (OO) infant; the same pattern is true for two heterozygous type B (BO) parents
   (2) Two type AB parents may produce a homozygous type A (AA), homozygous type B (BB), or type AB infant
   (3) A heterozygous type A (AO) parent and a type AB parent can produce a homozygous type A (AA), heterozygous type A (AO), heterozygous type B (BO), or type AB infant

C. Rh blood group system

1. In the Rh system, erythrocytes are classified according to a group of inherited antigens on their surface
2. Scientists performed the original Rh studies by immunizing rabbits and guinea pigs with blood from rhesus monkeys
   a. The immunized animals formed anti-Rhesus antibodies that reacted not only with the rhesus monkey erythrocytes but with the erythrocytes of about 85% of White individuals and 95% of Black individuals; those whose erythrocytes were agglutinated by the anti-Rhesus antibodies were called Rhesus-positive or simply Rh-positive
   b. Those whose erythrocytes did not react with the anti-Rhesus antibodies were called Rhesus-negative or Rh-negative
3. When exposed to Rh-positive blood by blood transfusion or pregnancy with an Rh-positive fetus, an Rh-negative individual may form anti-Rh antibodies
   a. Subsequent transfusion of Rh-positive blood to this individual leads to rapid destruction of the transfused cells (transfusion reaction)
   b. If a woman whose blood contains anti-Rh antibodies becomes pregnant with an Rh-positive fetus, the anti-Rh antibodies cross the placenta and destroy fetal erythrocytes, causing hemolytic disease of the newborn
4. Although many genes and erythrocyte antigens are involved in the Rh system, the Rh antigen described first is the most important clinically; it is called the D antigen or the Rh_o antigen (the subscript *o* stands for *original)*
5. The gene D and its allele d determine the presence of D antigen on erythrocytes
6. Individuals whose erythrocytes contain the D antigen (genotype DD or Dd) are considered Rh-positive (regardless of other Rh antigens on the erythrocytes); individuals who lack the D antigen (genotype dd) are considered Rh-negative

7. An Rh-positive individual may be homozygous (DD) or heterozygous (Dd); the latter is more common
8. Because each parent transmits one of two alleles to offspring, an infant may not have the same Rh genotype or phenotype as either parent
   a. Two heterozygous Rh-positive (Dd) parents may produce a homozygous Rh-positive (DD), heterozygous Rh-positive (Dd), or Rh-negative (dd) infant
   b. A homozygous Rh-positive (DD) parent and a heterozygous Rh-positive (Dd) parent may produce a homozygous Rh-positive (DD) or heterozygous Rh-positive (Dd) infant
   c. A heterozygous Rh-positive (Dd) parent and an Rh-negative (dd) parent may produce an Rh-positive (Dd) or Rh-negative (dd) infant

Study Activities

1. Discuss the major components of blood and explain their functions.
2. Trace the process of hematopoeisis from one stem cell to all seven of the body's blood cells.
3. Describe the importance of iron and hemaglobin to erythrocyte function.
4. Compare and contrast the five different types of leukocytes.
5. Discuss the role of the platelets in hemostasis.
6. Contrast the factors that activate coagulation with those that inhibit it.
7. Explain how fibrinolysis dissolves and removes clots.
8 Compare the ABO and Rh blood group systems.

8

Lymphatic System and Immunity

Objectives

After studying this chapter, the reader should be able to:

- Identify the major organs of the lymphatic system and describe their functions.
- Discuss the formation of the T and B lymphocytes.
- Contrast nonspecific resistance and acquired immunity.
- Discuss the major mechanisms of nonspecific resistance.
- Compare cell-mediated immunity and humoral immunity.
- Explain the importance of maintaining a proper balance between helper and suppressor lymphocytes.
- Describe the basic structure and function of the five types of immunoglobulins.
- Describe how IgE causes allergic manifestations in susceptible individuals.

I. Lymphatic System

A. General information

1. The *lymphatic system* consists of lymphocyte-containing (lymphoid) tissues and lymphatic vessels
2. Lymphoid tissues are concentrated in the lymph nodes, spleen, and thymus; they also are present in the mucous membranes of the respiratory and gastrointestinal tracts
3. Lymphoid tissues are connected by a complex network of lymphatic vessels, which are thin-walled drainage channels similar to veins
    a. These vessels collect *lymphatic fluid* or *lymph* (liquid that contains lymphocytes and is derived from tissue fluid) and return it to the circulation
    b. Small lymphatic channels converge to form larger vessels with valves that prevent reflux
4. Lymphatic vessels eventually empty into the right or left subclavian vein
    a. The right lymphatic duct empties lymph from the upper right part of the body into the right subclavian vein
    b. The thoracic duct empties lymph from the rest of the body into the left subclavian vein
5. The lymphatic system has three main functions
    a. Lymphatic vessels collect and return to the bloodstream some of the fluid filtered from the capillaries into the interstitial tissues (see Chapter 6, Cardiovascular System, for more information)

b. Lymphatic vessels absorb fats from the gastrointestinal tract and transport them into the bloodstream
c. Lymphatic tissue forms the cornerstone of the immune system

B. Lymphocytes

1. Lymphocytes are a type of leukocyte (white blood cell); they develop from stem cells in the bone marrow, which differentiate into lymphocyte precursor cells (see Chapter 7, Blood, for more information on blood cell development)
2. Some precursor cells migrate from the bone marrow to the thymus, where they undergo further differentiation and develop into *T (thymus-dependent) lymphocytes*
3. Other precursor cells continue to differentiate in the bone marrow, where they develop into *B (bone marrow-dependent) lymphocytes*
4. T and B lymphocytes migrate into the lymph nodes, spleen, and other lymphatic tissues, where they proliferate to form the mature lymphocytes that populate these lymphoid tissues
   a. These lymphocytes do not remain permanently in a specific lymphatic organ; they continually recirculate between the blood and various lymphatic tissues and organs
   b. About 70% of these circulating cells are T lymphocytes; most of the remainder are B lymphocytes
5. A small proportion of lymphocytes are called *null cells*
   a. Null cells cannot be classified as B or T lymphocytes
   b. They are hematopoietic stem cells and may include B- and T-lymphocyte precursors and precursor, myeloid, and platelet cells
   c. Nulls cells can destroy tumor cells spontaneously or through an antibody-dependent cellular cytotoxic mechanism

C. Lymph nodes

1. Lymph nodes are small bean-shaped structures located along lymphatic vessels, where they act as filters
2. Each lymph node consists of a mass of lymphocytes supported by a meshwork of reticular fibers; it is enclosed by a fibrous capsule and divided into an outer cortex and inner medulla
   a. The outer cortex divides into the superficial and deep cortex
      (1) The superficial cortex contains follicles composed predominantly of B cells; in an immune response, the follicles expand and develop clusters of proliferating cells, called germinal centers
      (2) The interfollicular areas and deep cortex contain mostly T cells
   b. The inner medulla contains loosely arranged lymphocytes and scattered groups of mononuclear phagocytes derived from monocytes (see Chapter 7, Blood, for more information on blood cell development)
3. Lymph flows into the lymph node through afferent lymphatic vessels into a subcapsular sinus at the periphery of the node; lymph filters slowly through the node and is collected into efferent lymphatic channels, which drain from an indentation on the concave side (hilus) of the node
   a. As lymph flows through the node, mononuclear phagocytes filter out and destroy foreign substances
   b. They also interact with lymphocytes to generate an immune response (discussed later in this chapter)

D. Spleen

1. The spleen is an ovoid, fist-sized organ that filters antigens and other particles from the blood
2. It is surrounded by a dense fibrous capsule from which bands of connective tissue extend into its interior, which is called the splenic pulp
3. The pulp is divided into white pulp and red pulp
   a. White pulp is composed of compact masses of lymphocytes that surround branches of the splenic artery
   b. Red pulp consists of a network of blood-filled sinusoids, which are supported by a framework of reticular fibers and star-shaped mononuclear phagocytes; lymphocytes, plasma cells, and monocytes also appear in the framework
      (1) In the red pulp, pulp cords surround and separate the sinusoids
      (2) Long, narrow endothelial cells line the splenic sinusoids; they lie parallel to the long axis of the sinusoids and are supported by a fenestrated basement membrane
4. Blood enters the spleen in two ways
   a. Splenic blood flows from branches of the splenic arteries; some of it flows directly into the splenic sinusoids
   b. Most of it is discharged from branches of the splenic arteries directly into the pulp cords
      (1) Blood cells pass from the meshwork of cells and fibers between the pulp cords into the sinusoid lumens by squeezing through the long, slitlike openings between adjacent endothelial cells
      (2) This forces the blood to flow through the framework of reticular fibers, macrophages, and other cells before entering the sinusoids
5. The spleen performs several functions
   a. Splenic phagocytes engulf and break down worn-out erythrocytes; this action releases hemoglobin, which is broken down into its components (see Chapter 7, Blood, for more information)
   b. Splenic phagocytes also selectively retain and destroy damaged or abnormal erythrocytes and cells with a large amount of abnormal hemoglobin
   c. The spleen filters out bacteria and other foreign substances that enter the blood stream; splenic phagocytes promptly remove these substances
   d. Phagocytes also interact with lymphocytes to initiate an immune response (see Section II, Immunity, for more information)
6. Injury or disease may require spleen removal, which affects the body's defense mechanisms
   a. Bacteria elimination and antibody production are less efficient
   b. Consequently, the individual becomes susceptible to serious blood infections caused by various pathogenic organisms

E. Thymus

1. The thymus is a bilobular mass of lymphoid tissue located over the base of the heart
2. This organ helps develop T lymphocytes in the fetus and in the infant for a few months after birth; it has no function in the body's immune defenses after this time
3. The thymus is a large structure in infants but gradually undergoes atrophy when its function is no longer required; only a remnant persists in adults

F. Other lymphatic tissues

1. Other lymphatic tissues include the tonsils, adenoids, appendix, and Peyer's patches in the intestines
2. These tissues are distributed in mucous membranes where they can intercept invading organisms or toxins before they can spread widely
   a. Tissues in the throat and pharynx (tonsils and adenoids) can intercept antigens that enter by the upper respiratory tract
   b. Tissues in the gastrointestinal tract (the appendix and Peyer's patches) can intercept antigens that attempt to enter via the gut

II. Immunity

A. General information

1. The body has two mechanisms to protect itself against microorganisms and other potentially harmful substances
2. The first is a group of general protective mechanisms that function without prior exposure to harmful agents; these mechanisms provide *nonspecific resistance*
3 The second mechanism depends on the lymphatic system and provides *acquired* ***immunity***
   a. Acquired immunity consists of specific immune responses directed against specific organisms or toxins
   b. These responses usually require previous exposure to a foreign substance and cause antibody formation or lymphatic activation

B. Nonspecific resistance

1. Five chief mechanisms provide nonspecific resistance
2. Intact skin resists invasion by organisms by preventing their attachment; skin desquamation and low pH further impede bacterial colonization
3. Mucus in the respiratory passages traps bacteria and other foreign material
4. Gastric acid secretions and digestive enzymes destroy organisms swallowed into the stomach
5. Chemical compounds in the blood attach to and destroy foreign organisms or toxins
   a. The enzyme lysozyme, which is present in tears, nasal secretions, perspiration, and saliva, acts as an antibacterial agent
   b. Basic polypeptides inactivate certain gram-positive bacteria
   c. The serum protein properdin destroys gram-negative bacteria
6. The *inflammatory response* (a nonspecific reaction to any harmful agent) mobilizes leukocytes to engulf and destroy bacteria and other foreign material
   a. After organisms invade, tissue injury leads to release of histamine, kinins, and prostaglandins, which cause vasodilation and increased capillary permeability
   b. This increases blood flow to the affected tissues, where fluid collects
   c. Neutrophils and other leukocytes are attracted to the invasion site
   d. These cells engulf and destroy the organisms, foreign substances, and debris

C. Acquired immunity

1. Acquired immunity consists of two mechanisms: cell-mediated immunity and humoral immunity
2. Both mechanisms provide immunity against specific invading agents, such as viruses, bacteria, toxins, and other foreign substances; they are discussed in detail later in this chapter
   a. Cell-mediated immune functions are carried out by T lymphocytes
      (1) They require formation of large numbers of sensitized lymphocytes
      (2) After phagocytic cells (macrophages) present processed antigens to the lymphocytes, the T lymphocytes are sensitized by the antigens and become capable of destroying them on subsequent exposure
   b. Humoral immune functions are carried out by B lymphocytes
      (1) They require formation of *antibodies* (immunoglobulin formed in response to a specific antigen)
      (2) After macrophages present processed antigens to the lymphocytes, the B lymphocytes differentiate into plasma cells that produce antibodies
   c. Both types of immunity interact to protect the body against invading antigens
      (1) Some B and T lymphocytes retain a memory of the sensitizing antigen, which they pass on to succeeding generations of lymphocytes
      (2) Later contact with this antigen leads to rapid proliferation of sensitized lymphocytes or antibody-forming plasma cells
3. Acquired immunity develops after the first invasion by a foreign organism or first contact with a toxin
   a. Each toxin or type of organism contains specific chemical compounds that make it different from all other substances
   b. These compounds, called *antigens,* cause acquired immunity; they usually are high-molecular-weight proteins, polysaccharides, or lipids
4. The initial phase of the immune response involves macrophage and lymphocyte interaction; both types of cells are distributed widely throughout the body and can respond to foreign material wherever they encounter it
   a. Macrophages ingest the foreign material, process its antigens, and present the processed material the lymphocytes
   b. The lymphocytes respond and transform into antibody-forming plasma cells and sensitized lymphocytes, respectively, which proliferate rapidly to perform immune functions
5. After initial contact with an antigen, several weeks are required for macrophages to process the antigen and for lymphocytes to respond
6. Once the body has reacted to the antigen, some lymphocytes (called *memory cells*) retain the ability to respond promptly to the same antigen on subsequent exposure; successive generations of lymphocytes derived from these memory cells also retain this ability
   a. Subsequent contact with the sensitizing antigen provokes a renewed immune response
   b. This response is rapid because earlier exposure to the antigen has primed the immune system
7. The ability to generate an immune response is controlled genetically; immune-response genes regulate T- and B-lymphocyte proliferation

III. Cell-Mediated Immunity

A. General information

1. Cell-mediated immunity results from the function of sensitized T lymphocytes
2. This mechanism is the main defense against viruses, fungi, parasites, and some bacteria
3. Cell-mediated immunity also eliminates abnormal cells that may arise during cell division; these cells can develop into tumors if not destroyed
4. This mechanism also causes organ transplant rejection
5. Cell-mediated immunity commonly is associated with hypersensitivity to bacterial antigens or other foreign substances; this reaction is characterized by an intense inflammatory reaction at the site of contact with the foreign material

B. Cell-mediated immune response

1. Three types of sensitized T lymphocytes perform specific functions in cell-mediated immune response: *cytotoxic T cells, helper T cells,* and *suppressor T cells*
2. Also called *killer cells,* cytotoxic T cells directly kill organisms or other invading cells
   a. Cytotoxic T cells bind tightly to the invading cell, swell, and release cytotoxic substances directly into the attacked cell
   b. These cells can attack and kill many organisms in succession, without being harmed
3. Helper T cells interact with other T and B lymphocytes to enhance the immune response
   a. When activated, helper T cells secrete *lymphokines,* which increase activation of B cells, cytotoxic T cells, and suppressor T cells by antigens
   b. They attract macrophages and promote more efficient phagocytosis
4. Suppressor T cells inhibit the immune response
   a. They suppress the actions of cytotoxic and helper T cells
   b. This regulates the other T cells, preventing them from causing excessive immune reactions and severe tissue damage
5. Sensitized T lymphocytes are classified into two major groups according to specific antigens on their membranes
   a. One group, the CD_4 lymphocytes, are the helper T cells; they account for about 70% of the T cells
   b. The other group, the CD_8 lymphocytes, consists of the suppressor and cytotoxic T cells; they account for about 30% of all T cells
   c. A normal ratio of CD_4 to CD_8 lymphocytes (which reflects the normal proportions of helper to suppressor and cytotoxic T cells) is essential for proper immune function
      (1) Loss or destruction of helper T cells (as in acquired immunodeficiency syndrome) leads to a relative excess of suppressor T cells, inhibiting the immune response and increasing susceptibility to infection
      (2) A relative lack of suppressor T cells allows the immune system to respond unchecked, increasing the likelihood of autoimmune disease in which the immune defenses attack the body's own cells and tissues

IV. Humoral Immunity

A. General information

1. Humoral immunity results from B lymphocyte function
2. It forms the major defense against many bacteria and bacterial toxins
3. In response to antigen stimulation, B lymphocytes differentiate into plasma cells that produce antibodies (immunoglobulins) that can combine with and eliminate the foreign substance

B. Antibody mechanisms of action

1. Antibodies act by direct attack on the invader or by activating the ***complement system,*** which destroys the invader
2. Antibodies can directly inactivate an invader in one of four ways
   a. Through *agglutination,* antibodies can cause clumping of multiple large structures with antigens on their surfaces, such as erythrocytes or bacteria
   b. Through *precipitation,* antibodies produce an antigen-antibody complex so large that it is insoluble and precipitates; tetanus toxin is a soluble antigen that is susceptible to precipitation
   c. Through *neutralization,* antibodies cover the toxic sites on an antigen
   d. Through *lysis,* potent antibodies attack the cell membrane of an antigen, causing it to rupture; this process requires complement activation
3. The complement system, which includes a group of proteins in blood and tissue fluids, augments the effects of the direct actions of antibodies
   a. Exposure to an antigen activates the complement system, triggering a complex cascade of sequential reactions
   b. The cascade ultimately produces many end-products
   c. Some of the end-products help prevent antigen-induced damage by causing certain effects
      (1) Some end-products increase phagocytosis through ***opsonization,*** which can increase bacteria destruction by many hundredfold
      (2) The end-product called lytic complex directly causes cell *lysis* by rupturing the membrane of bacteria and other organisms
      (3) Some end-products cause invading organisms to adhere to each other by altering their surfaces; this produces agglutination
      (4) Enzymes and other end-products can neutralize viruses by attacking their structures
      (5) By causing ***chemotaxis*** (movement toward a chemical stimulus), one end-product attracts many neutrophils and macrophages to the area affected by the antigen

C. Immunoglobulins

1. Immunoglobulins are proteins produced by the plasma cells derived from B lymphocytes
2. Immunoglobulins differ in chemical composition, molecular weight, and size, but they all have the same basic structure: two matched pairs of polypeptide (protein) chains bound together by chemical bonds
3. The chains are arranged to resemble a fork with four tines
   a. The two central chains, called heavy chains, form the handle and inner tines of the fork
   b. The two outer chains, called light chains, form the outer tines of the fork

c. The open end of each tine is different for each immunoglobulin; this variable part gives the immunoglobulin its specificity (ability to respond to a particular antigen)
d. The handle end is the constant part of the immunoglobulin; it is the same in every immunoglobulin
e. The constant part does not combine with an antigen; however, it determines other properties of the immunoglobulin, such as its ability to activate complement and attach to the surface of an antigen's cell membrane

4. All immunoglobulins are composed of the same basic four-chain units, but some aggregate to form clusters of two, three, or five units
5. Five different types of antibodies exist: immunoglobulin M (IgM), immunoglobulin G (IgG), immunoglobulin A (IgA), immunoglobulin D (IgD), and immunoglobulin E (IgE)
   a. IgM is the largest antibody molecule
      (1) Each molecule is an aggregate of five basic immunoglobulin units
      (2) It commonly is called a macroglobulin because of its large size and high molecular weight
      (3) It is particularly efficient in combining with large particulate antigens
   b. IgG is the principal immunoglobulin molecule formed in response to most infectious agents
      (1) Each molecule is composed of a single basic immunoglobulin unit
      (2) It combines with antigens and activates complement by functioning as an opsonin; it also causes antibody activity in tissues
   c. IgA is produced by antibody-forming cells in the respiratory system, gastrointestinal tract, and other mucous membranes
      (1) It appears in secretions
      (2) IgA is an aggregate of two basic immunoglobulin units
      (3) It combines with potentially harmful ingested or inhaled antigens, preventing their absorption
   d. IgD coats the surface of B lymphocytes
      (1) It is composed of one basic immunoglobulin unit
      (2) Its prongs (antigen-binding sites) can attach antigens to the surface of B lymphocytes stimulating them to produce antibodies
   e. IgE usually is present in small amounts in the blood of most persons; it is composed of a single immunoglobulin unit and appears in much larger amounts in the blood of allergic individuals
      (1) Some individuals form specific IgE antibodies (become allergic) to ragweed, plant pollens, and other substances that do not affect most people; the allergy-prone individual is called *atopic*, and the sensitizing antigen is called an *allergen*
      (2) The handle end of IgE attaches to the membrane of mast cells and basophils; the tine ends project from the cell membrane
      (3) Subsequent contact with the sensitizing antigen causes the antigen to fix to the antigen-combining sites on the projecting ends of the IgE molecules
         (a) This event triggers basophil and mast cell granules to release histamine and other chemicals; because these chemicals incite an inflammatory reaction, they are called mediators of inflammation

Allergic Response

An allergen is an antigen that triggers an allergic response. The first exposure to an allergen causes lymphocytes and plasma cells to produce specific immunoglobulin E (IgE) antibodies, which bind to mast cells and basophils. Subsequent exposure to this allergen leads to antigen-antibody interaction, causing mast cells and basophils to release histamine and other mediators. These mediators produce allergic manifestations.

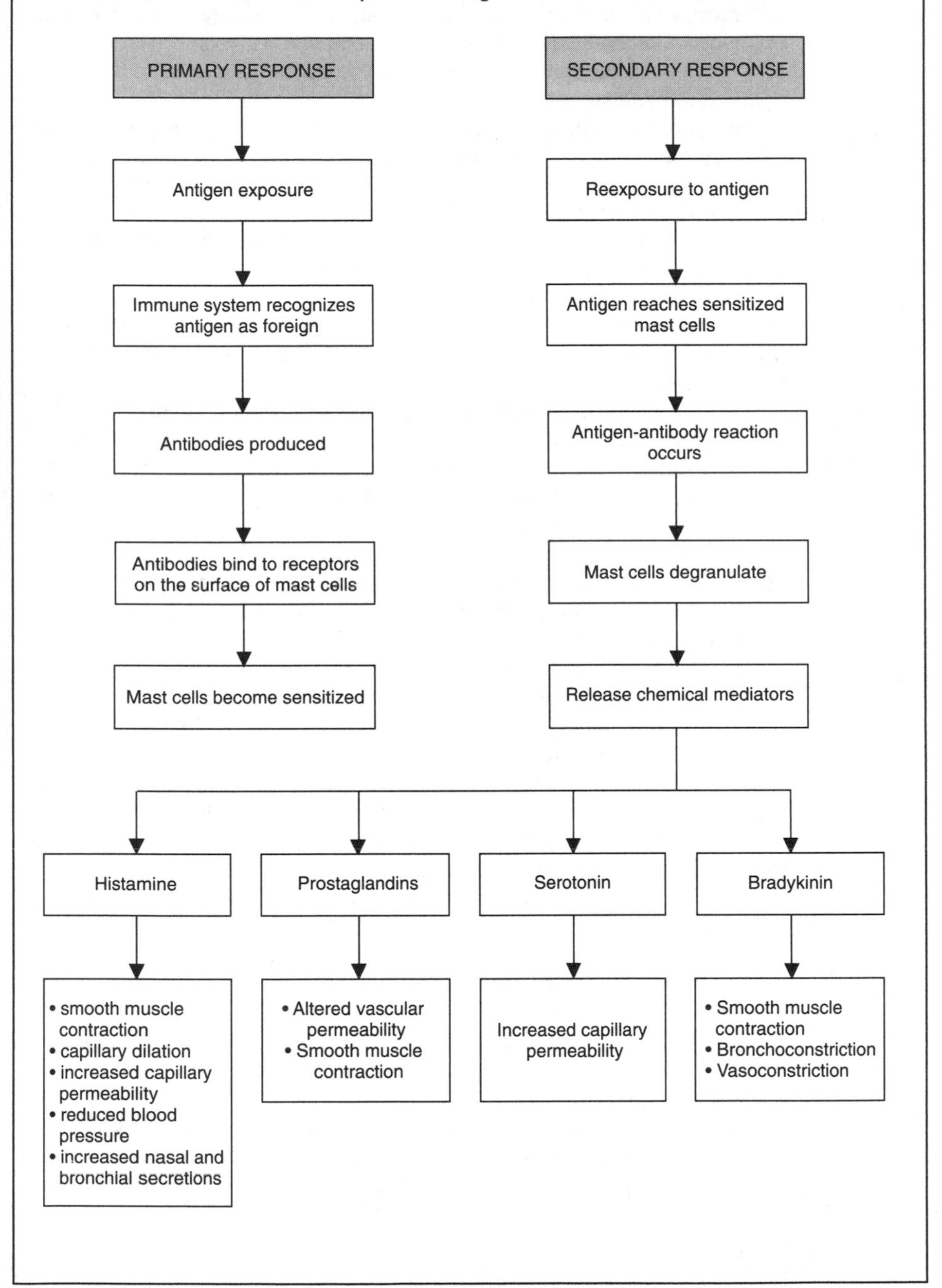

(b) These chemical mediators produce allergic manifestations, such as sneezing, stuffy nose, and itchy eyes (see *Allergic Response* for a diagram)

Study Activities

1. Locate the major lymphatic tissues and organs and explain their functions.
2. Compare the formation and functions of T and B lymphocytes.
3. Explain how the body maintains nonspecific resistance.
4. Explain how the body obtains acquired immunity.
5. Describe the role of T lymphocytes in cell-mediated immunity.
6. Discuss the types and actions of antibodies in humoral immunity.
7. Identify the causes and effects of an allergic response.

9

Nutrition, Digestion, and Metabolism

Objectives

After studying this chapter, the reader should be able to:
- Describe the major types of nutrients and their chemical structures.
- Discuss the function of each gastrointestinal (GI) tract structure.
- Explain how GI hormones regulate gastrointestinal, pancreatic, and biliary functions.
- Compare the functions of the exocrine and endocrine pancreas.
- Describe the metabolic functions of the liver.
- Discuss the composition and functions of bile.
- Explain how carbohydrates, proteins, and lipids are broken down, absorbed, and metabolized.
- Describe the three processes by which glucose is broken down to yield energy.
- Discuss the role of the liver in regulating blood glucose and muscle cell glycogen.
- Define deamination and transamination and explain their purpose in protein metabolism.
- Describe the significance of ketone bodies in lipid metabolism.
- Compare the functions of the six hormones that regulate metabolic processes.

I. Nutrition

A. General information

1. The body needs a continual supply of water and various nutrients to maintain its functions; virtually all nutrients come from digested food
2. The three major types of nutrients the body needs are *carbohydrates, proteins,* and *lipids*
   a. When nutrients are used to yield energy, the energy is measured in kilocalories (kcal), or calories, per gram of nutrient
   b. Adults require approximately 2,000 kcal daily
3. *Vitamins* are essential for normal ***metabolism;*** these compounds contribute to enzyme reactions that promote metabolism of carbohydrates, proteins, and lipids
4. *Minerals* participate in enzyme metabolism, membrane transfer of essential elements, and other vital functions

B. Carbohydrates

1. Carbohydrates are organic compounds composed of carbon, hydrogen, and oxygen
2. They yield 4 kcal/g when used for energy

3. Carbohydrates are classified as monosaccharides, disaccharides, and polysaccharides
4. A monosaccharide is a simple sugar that cannot be decomposed by ***hydrolysis***
   a. A monosaccharide is a polyhydroxy aldehyde or ketone
      (1) An *aldehyde* is a substance that contains the characteristic group -CHO; the term *polyhydroxy* means that the carbon atoms are linked to a hydroxyl (OH) group
      (2) A ketone is a substance that contains the carbonyl group CO and carbon groups attached to the carbonyl carbon
   b. Monosaccharides are classified by the number of carbon atoms they contain
      (1) A pentose is a five-carbon monosaccharide, such as ribose in ribonucleic acid (RNA) and deoxyribose in deoxyribonucleic acid (DNA)
      (2) A hexose is a six-carbon compound, such as glucose, fructose, and galactose
      (3) Other monosaccharides include diose (two-carbon), triose (three-carbon), tetrose (four-carbon), and heptose (seven-carbon) monosaccharides
   c. Monosaccharides have a ringlike (cyclic) structure because of the chemical attraction between the aldehyde or ketone at one end of the chain and the hydroxyl group at the other end
5. In a disaccharide, the hydroxyl groups of two monosaccharides join end-to-end; in this ***condensation reaction,*** the monosaccharides release a molecule of water (H_2O), and a glycoside bond forms
   a. Joining of a glucose and a fructose molecule yields the disaccharide sucrose
   b. Joining of a glucose and a galactose molecule yields the disaccharide lactose
6. A polysaccharide consists of a long chain (polymer) of more than 10 monosaccharides linked by glycoside bonds; the polysaccharide starch is a polymer of glucose

C. Proteins

1. Proteins are complex nitrogenous organic compounds that contain amino acid chains; some also contain sulfur and phosphorus
2. Proteins yield 4 kcal/g when used for energy, although they are used primarily for growth and repair of body tissues
3. Some proteins are combined with lipids (lipoproteins) or carbohydrates (glycoproteins)
4. All *amino acids* have some structural similarities
   a. They all contain a carbon atom to which a carboxyl group (COOH) and an amino group (NH_2) are attached
   b. They also contain a side chain (symbolized by the letter R), which is where amino acids show structural differences
5. Amino acids are joined by condensation of the carboxyl group on one amino acid with the amino group of the adjacent amino acid
   a. The reaction releases a molecule of water
   b. It produces a linkage called a *peptide bond*
6. A chain of 2 to 10 amino acids forms a *peptide;* a chain of 10 or more amino acids forms a *polypeptide*
7. A chain of more than 50 amino acids forms a protein

a. The sequence and types of amino acids in the chain determine the character of the protein
b. Each protein is synthesized on a ribosome as a straight chain
c. Chemical attractions between the amino acids in various parts of the chain cause it to coil or twist into the specific configuration of each protein

D. Lipids

1. Lipids are organic compounds that are insoluble in water but soluble in alcohol and other organic solvents
2. A concentrated form of fuel, they yield approximately 9 kcal/g when used for energy
3. The major lipids include fats (the most common), phospholipids, and steroids
4. A fat (triglyceride) is composed of three molecules of fatty acid combined with one molecule of glycerol
   a. A fatty acid is a long-chain compound with an even number of carbon atoms (usually 16 or 18) and a terminal carboxyl group
   b. Glycerol is a three-carbon compound (alcohol) with a hydroxyl (OH) group attached to each carbon atom
      (1) The carboxyl group on each fatty acid molecule combines with one hydroxyl group on the glycerol molecule, releasing a molecule of water
      (2) The carboxyl group –hydroxyl group linkage produces an ester linkage
   c. Fats can be classified by the number of double bonds between carbon atoms in the fatty acid molecules
      (1) Saturated fats have no double bonds between carbon atoms
      (2) Monounsaturated fats contain one double bond between carbon atoms
      (3) Polyunsaturated fats have multiple double bonds between carbon atoms
5. Phospholipids are similar to fats, but a phosphorus- and nitrogen-containing compound replaces one of the fatty acid molecules
6. Steroids, such as cholesterol, bile salts, and sex hormones, are complex molecules in which the carbon atoms form four cyclic structures, to which various side chains are attached; they contain no glycerol or fatty acid molecules

E. Vitamins and minerals

1. Vitamins are biologically active organic compounds that are essential for normal metabolism and growth and development
2. Vitamins are classified as water-soluble or fat-soluble
   a. Water-soluble vitamins include the B complex and C vitamins
   b. Fat-soluble vitamins include vitamins A, D, E, and K
3. Minerals are inorganic substances that are required for enzyme metabolism, membrane transfer of essential compounds, maintenance of acid-base balance and osmotic pressure, nerve impulse transmission, muscle contractility, and growth
4. Minerals appear in bones, hemoglobin, thyroxin, and vitamin B_{12}; they may be classified as major minerals (more than 0.005% of body weight) or trace minerals (less than 0.005% of body weight) (see *Vitamin and Mineral Functions* for more information about these nutrients)
   a. Major minerals include calcium, chloride, magnesium, phosphorus, potassium, sodium, and sulfur

Vitamin and Mineral Functions

Water-soluble and fat-soluble vitamins and major and trace minerals perform vital functions in the body, as shown in the chart below.

NUTRIENT	MAJOR FUNCTIONS
Water-soluble vitamins	
Vitamin C (ascorbic acid)	• Collagen formation • Bone and tooth formation • Iodine conservation • Healing • Erythrocyte formation • Infection resistance • Iron absorption and use • Corticosteroid synthesis
Vitamin B_1 (thiamine)	• Carbohydrate, fat, and protein metabolism • Energy production • Central nervous system (CNS) maintenance
Vitamin B_2 (riboflavin)	• Antibody and erythrocyte formation • Energy production • Epithelial, eye, and mucosal tissue maintenance
Vitamin B_6 (pyridoxine)	• Antibody formation • Digestion • Deoxyribonucleic acid (DNA) and ribonucleic acid (RNA) synthesis • Fat and protein metabolism • Hemoglobin production • Sodium and potassium balance • CNS maintenance • Tryptophan to niacin conversion
Folic acid (folacin, pteroylglutamic acid)	• Erythrocyte and leukocyte formation and maturation • DNA and RNA formation
Niacin (nicotinic acid, nicotinamide, niacinamide)	• Cholesterol level reduction • Carbohydrate, protein, and fat metabolism • Sex hormone production • Glycogen synthesis
Vitamin B_{12} (cyanocobalamin)	• Erythrocyte maturation • Cellular and nutrient metabolism • Cell longevity • Iron absorption • Tissue growth • Nerve cell maintenance • Myelin formation
Biotin	• Cell growth • Fatty acid synthesis • Carbohydrate, fat, and protein metabolism • Vitamin B use • Energy production
Pantothenic acis (formerly vitamin B_3)	• Antibody formation • Carbohydrate, fat, and protein metabolism • Cortisone production • Growth stimulation • Stress tolerance • Cholesterol synthesis
Fat-soluble vitamins	
Vitamin A (retinol, provitamin A)	• Body tissue repair and maintenance • Infection resistance • Bone growth • Nervous system development • Cell membrane metabolism and structure • RNA synthesis • Visual purple production (for night vision)

(continued)

Vitamin and Mineral Functions *(continued)*

NUTRIENT	MAJOR FUNCTIONS
Fat-soluble vitamins *(continued)*	
Vitamin D (calciferol; subtypes include D_2 [ergocalciferol] and D_3 [cholecalciferol])	• Calcium and phosphorus use • Mineralization of bones and teeth • Serum calcium level regulation
Vitamin E (tocopherol)	• Cell membrane protection • Erythrocyte hemolysis prevention • Sexual potency and fertility maintenance (unproved)
Vitamin K (menadione; subtypes include K_2 [menaquinone] and K_3)	• Liver synthesis of prothrombin and other blood clotting factors
Major minerals	
Calcium	• Blood clotting • Bone and tooth formation • Cardiac rhythm regulation • Cell membrane structure and function • Muscle growth and contraction • Nerve impulse transmission
Chloride	• Maintenance of fluid, electrolyte, acid-base, and osmotic pressure balance
Magnesium	• Acid-base balance • Calcium and phosphorus metabolism in bones • Muscle relaxation • Cellular respiration • Nerve impulse transmission • Cardiac muscle function and maintenance
Phosphorus	• Bone and tooth formation • Cell growth and repair • Energy production • Kidney function • Carbohydrate, fat, and protein metabolism • Myocardial contraction • Nerve and muscle activity • Acid-base balance
Potassium	• Heartbeat regulation • Muscle contraction • Nerve impulse transmission • Fluid distribution and osmotic pressure balance • Acid-base balance
Sodium	• Extracellular fluid regulation • Osmotic pressure balance • Muscle contraction • Acid-base and water balance • Cell permeability • Muscle function • Nerve impulse transmission
Sulfur	• Collagen synthesis • Vitamin B formation • Muscle metabolism • Toxin neutralization • Blood clotting
Trace minerals	
Chromium	• Carbohydrate and lipid metabolism • Serum glucose level maintenance
Cobalt	• Vitamin B_{12} formation

Vitamin and Mineral Functions *(continued)*

NUTRIENT	MAJOR FUNCTIONS
Trace minerals *(continued)*	
Copper	• Bone formation • Healing processes • Hemoglobin, erythrocyte, and enzyme formation • Mental processes • Iron use
Fluoride (fluorine)	• Bone and tooth formation
Iodine	• Regulation of basal metabolic rate • Cell metabolism
Iron	• Growth (in children) • Hemoglobin production • Stress and disease resistance • Cellular respiration • Oxygen transport • Energy production • Regulation of biological and chemical reactions
Manganese	• Enzyme activation • Carbohydrate, fat, and protein metabolism • Skeletal growth • Sex hormone production • Vitamin B_1 metabolism • Vitamin E use
Molybdenum	• Body metabolism
Selenium	• Immune mechanisms • Mitochondrial adenosine triphosphate synthesis • Cellular protection • Fat metabolism
Zinc	• Burn and wound healing • Carbohydrate digestion • Carbohydrate, fat, and protein metabolism • Prostate gland function • Reproductive organ growth and development • Taste and smell regulation

Adapted from Quillman, S. (1990). *Nutrition and Diet Therapy.* Springhouse Notes. Springhouse, PA: Springhouse Corp.

b. Trace minerals include chromium, cobalt, copper, fluorine, iodine, iron, manganese, molybdenum, selenium, and zinc

II. Digestive System Structures and Functions

A. General information

1. The digestive system consists of the gastrointestinal (GI) tract (sometimes called the alimentary canal) and accessory organs
2. In the digestive system, several organs help digest, or break down, and process food

B. GI tract

1. The GI tract is a tube in which food is digested and absorbed; it stretches from the mouth to the anus
2. It consists of the oral cavity, pharynx, esophagus, stomach, small intestine, and large intestine
3. The oral cavity prepares food for swallowing and begins ***digestion***
   a. The teeth cut and grind food into small particles
   b. The salivary glands pour saliva into the mouth
      (1) The food mixes with saliva to form a pliable mass (bolus) for swallowing
      (2) Saliva contains amylase, an enzyme that begins starch digestion
4. Swallowing propels the food bolus from the mouth into the pharynx, which runs from the level of the base of the skull to the sixth vertebra
5. Reflex contractions of pharyngeal muscles move the bolus into the esophagus, which extends from the pharynx to the stomach
   a. The upper esophageal sphincter relaxes, and the bolus is transported down the esophagus by ***peristalsis*** through the wavelike, rhythmic contractions of smooth muscle
   b. The lower esophageal sphincter relaxes, allowing the bolus to pass into the stomach
6. The stomach churns the ingested food and mixes it with gastric juices; this results in a thick, acidic fluid called *chyme*
   a. GI hormones, such as gastrin, gastric inhibitory peptides, secretin, and cholecystokinin, regulate gastric secretion and motility
   b. Stimulation of gastric acid secretion occurs during two phases
      (1) In the cephalic phase, secretion is stimulated by the sight, smell, or anticipation of food
      (2) In the gastric phase, secretion is stimulated by *gastrin,* the GI hormone released in response to food in the stomach and duodenum
   c. Gastric acid activates the gastric enzyme precursor pepsinogen to form pepsin, which digests proteins by breaking peptide bonds
   d. Intrinsic factor (IF) in gastric acid combines with vitamin B_{12}; the vitamin B_{12}-IF complex is absorbed later in the small intestine
7. The small intestine is composed of the duodenum, ileum, and jejunum; the stomach expels chyme through the pylorus into the upper part of the small intestine (duodenum)
   a. Segmenting contractions (alternating contractions and relaxations of adjacent segements of the small intestine) mix the contents of the small intestine, which are propelled by peristalsis
   b. Most of the nutrients, water, and electrolytes in foods are digested and absorbed during the 6- to 8-hour passage through the small intestine
   c. Intestinal glands secrete enzymes that further digest various nutrients
   d. The intestinal hormones cholecystokinin and secretin also regulate gallbladder function and pancreatic fluid and bile secretion (see *Gastrointestinal Hormone Production and Function* for more information)
   e. Bile and pancreatic fluid mix with intestinal contents to continue digestion, as discussed in the next section
8. The large intestine consists of the colon and rectum and extends from the small intestine to the anus

Gastrointestinal Hormone Production and Function

When stimulated by specific agents, gastrointestinal (GI) structures secrete four hormones that play different roles in digestion, as shown in the chart below.

HORMONE	PRODUCTION SITE	STIMULATING FACTOR OR AGENT	FUNCTION
Gastrin	Pyloric antrum and duodenal mucosa	• Pyloric antrum distention • Vagal stimulation • Protein digestion products • Alcohol	Stimulates gastric acid secretion and gastric motility
Gastric inhibitory peptides	Duodenal and jejunal mucosa	• Gastric acid • Fats • Fat digestion products	Stimulates secretion of bile and alkaline pancreatic fluid
Secretin	Duodenal and jejunal mucosa	• Gastric acid • Fat digestion products • Protein digestion products	Stimulates secretion of bile and alkaline pancreatic fluid
Cholecystokinin	Duodenal and jejunal mucosa	• Fat digestion products • Protein digestion products	Stimulates gallbladder contraction and secretion of enzyme-rich pancreatic fluid

a. In the large intestine, peristalsis and segmenting contractions move the intestinal contents slowly; water are minerals are absorbed from the intestinal contents, leaving a residue of fecal material
   (1) Intestinal bacteria act on the residue, releasing decomposition products and intestinal gases
   (2) These bacteria also synthesize vitamin K and some B vitamins, which are absorbed in the colon
b. Peristalsis propels fecal material into the rectum; this causes reflex contraction of rectal smooth muscle and relaxation of the internal anal sphincter (defecation reflex)
c. Voluntary relaxation of the external anal sphincter combined with bearing-down efforts results in evacuation of rectal contents

C. Accessory organs

1. Accessory organs include the pancreas, liver, biliary duct system, and gallbladder; these organs are located outside the GI tract, but their secretions are vital to digestion
2. The exocrine pancreas contains cells that secrete digestive enzymes into the duodenum through the pancreatic duct; the endocrine pancreas contains cells that secrete hormones directly into the bloodstream
   a. The exocrine pancreas secretes about 1,500 ml of alkaline pancreatic fluid into the duodenum daily
      (1) Pancreatic fluid contains potent digestive enzymes
         (a) Amylase digests starch
         (b) Trypsin, chymotrypsin, and carboxypeptidase digest proteins
         (c) Lipase digests certain lipids
         (d) Cholesterol esterase digests cholesterol esters
         (e) Ribonuclease and deoxyribonuclease digest nucleic acids

(2) The pancreas secretes proteolytic (protein-digesting) enzymes as inactive precursors
(a) The intestinal enzyme enterokinase activates trypsinogen to trypsin
(b) In turn, trypsin activates chymotrypsinogen and procarboxypeptidase to form chymotrypsin and carboxypeptidase
(3) The small intestine releases the hormone secretin when acidic chyme is expelled into the duodenum; this hormone stimulates the pancreas to secrete a large volume of pancreatic fluid that is low in enzymes
(4) The small intestine also releases cholecystokinin when lipids and proteins enter the duodenum; this stimulates secretion of pancreatic fluid, which is rich in digestive enzymes, and stimulates gallbladder contraction

b. The endocrine pancreas consists of about 1 million small clusters of endocrine cells (islets of Langerhans), each composed of three major cell types
(1) Alpha cells secrete glucagon to increase blood glucose in response to decreased blood glucose levels
(2) Beta cells secrete insulin to lower blood glucose in response to increased blood glucose levels; these cells are absent or defective in individuals with diabetes mellitus, resulting in elevated blood glucose
(3) Delta cells secrete somatostatin (growth hormone inhibitory hormone), which inhibits glucagon and insulin secretion (see Chapter 12, Endocrine System, for more information about the endocrine pancreas)

3. The liver receives the products of digestion through the portal vein
a. It can convert the absorbed hexoses, amino acids, and lipid digestion products into whatever nutrient mixture the body needs for metabolic processes
b. The liver also forms ketone bodies from products of lipid metabolism
c. The liver is essential for the synthesis and storage of several substances
(1) It produces blood proteins (such as albumin and globulin), lipoproteins, and proteins involved with blood coagulation
(2) It also stores a small reserve of fat and glycogen, iron for hemoglobin formation, and vitamins A, B_{12}, D, E, and K
d. The liver secretes 500 to 1,000 ml of bile daily to promote fat digestion (see *Functions of Digestive Secretions* for more information)
(1) Bile is a complex secretion composed of cholesterol, lecithin (a phospholipid), bile salts (composed of cholesterol and amino acid derivatives), minerals, bile pigments (derived from hemoglobin breakdown), and water
(2) Bile contains no digestive enzymes
(a) It functions as a detergent, emulsifying fats into small globules for more efficient digestion
(b) Bile salts form aggregates called *micelles,* which promote the absorption of fats and fat-soluble vitamins
e. The liver detoxifies or excretes many wastes and toxins
(1) It converts potentially toxic compounds, such as ammonia, to nontoxic compounds
(2) It inactivates such substances as drugs, antibiotics, and steroid hormones
(3) It excretes bilirubin derived from erythrocyte breakdown

Functions of Digestive Secretions

Most digestive secretions contain enzymes that have specific digestive functions, as shown below. Only one secretion (bile) performs its functions without enzymes.

DIGESTIVE SECRETION	ENZYME	FUNCTION
Saliva	• Amylase	• Digests starch
Gastric acid	• Pepsin • Intrinsic factor	• Digests protein • Promotes vitamin B_{12} absorption
Small intestine secretions	• Disaccharidases (such as maltase, sucrase, and lactase) • Peptidases • Amylase • Lipase • Enterokinase	• Digest carbohydrates • Digest proteins • Digests starch further • Digests fats • Activates pancreatic trypsinogen to form trypsin, which continues protein digestion
Bile	• None	• Emulsifies fats and other lipids
Pancreatic secretions	• Trypsin, chymotrypsin, and carboxypeptidase • Ribonuclease and deoxyribonuclease • Amylase • Lipase • Cholesterol esterase	• Digest protein • Digest nucleic acids • Digests starch further • Digests fats and other lipids • Splits cholesterol esters to cholesterol and fatty acids

(a) Primarily in the spleen, the cells of the mononuclear phagocyte (reticuloendothelial) system break down worn-out erythrocytes into their constituent parts: heme (the iron-porphyrin ring) and globin (a protein)

(b) The globin is broken down into amino acids, which are reused to make other proteins

(c) The iron is removed from the porphyrin ring of heme and reused to make more hemoglobin

(d) The iron-free porphyrin ring breaks open and forms free bilirubin, which is relatively insoluble; it is transported to the liver loosely bound to albumin

(e) Liver cells extract free bilirubin and enzymatically join (conjugate) the bilirubin with glucuronic acid to form the more soluble conjugated bilirubin, which is excreted in bile

(f) In the colon, intestinal bacteria break down bilirubin into various compounds that produce the normal color of feces

(4) The liver also excretes cholesterol and other inactivated or detoxified products

4. Bile secreted from the liver drains into the biliary duct system through the hepatic ducts
   a. Bile entering the biliary excretory ducts is diverted into the cystic duct and enters the gallbladder, where it is stored and concentrated tenfold through water absorption
   b. Bile can enter the duodenum only when the ampullary sphincter of Oddi is open, during digestion

c. A fatty meal induces cholecystokinin release
   (1) This stimulates gallbladder contraction
   (2) Simultaneously, the ampullary sphincter relaxes, permitting bile to enter the small intestine
d. Various factors regulate cholesterol solubility in bile
   (1) The micelles formed from bile salts have a lipid-soluble center and a water-soluble periphery
   (2) Cholesterol remains in solution in bile by dissolving in the lipid-soluble centers of the micelles
   (3) Cholesterol may precipitate from bile and form gallstones if the cholesterol concentration in bile exceeds the capacity of the micelles to hold cholesterol in solution

III. Nutrient Digestion and Absorption

A. General information

1. All nutrients must be digested in the GI tract by enzymes that hydrolyze (split) large units into smaller ones
2. Then these smaller units are absorbed from the small intestine and transported to the liver through the portal venous system

B. Carbohydrate digestion and absorption

1. Enzymes break down complex carbohydrates into hexoses by hydrolyzing the glycoside bonds; hydrolysis restores the water molecules that were released when the bonds were formed
2. Salivary amylase begins starch hydrolysis into disaccharides in the oral cavity; pancreatic amylase continues this process in the small intestine
3. Intestinal mucosal disaccharidases hydrolyze disaccharides into monosaccharides
   a. Lactase hydrolyzes lactose to glucose and galactose
   b. Sucrase splits sucrose into glucose and fructose
4. Monosaccharides, such as glucose, fructose, and galactose, are absorbed through the intestinal mucosa by diffusion and active transport and are transported to the liver through the portal venous system
   a. Enzymes in the liver convert fructose and galactose to glucose
   b. Ribonucleases and deoxyribonucleases break down nucleotides from DNA and RNA into pentoses and nitrogen-containing compounds (nitrogen bases), which are absorbed through the intestinal mucosa like glucose

C. Protein digestion and absorption

1. Enzymes digest proteins by hydrolyzing peptide bonds
   a. These bonds link the amino acids that make up the protein chains
   b. This process restores water molecules that were released when the peptide bonds were formed
2. Gastric pepsin breaks proteins into polypeptides
3. Pancreatic trypsin, chymotrypsin, and carboxypeptidase convert polypeptides to peptides
4. Intestinal mucosal peptidases break peptides into their constituent amino acids

a. These amino acids are absorbed through the intestinal mucosa by active transport mechanisms
b. Then they are carried through the portal venous system to the liver, which converts amino acids not needed for protein synthesis into glucose

D. Lipid digestion and absorption

1. Bile from the liver emulsifies fats and other lipids into small droplets for more efficient digestion and eventual absorption in the small intestine
2. Pancreatic lipase breaks fats and phospholipids into a mixture of glycerol, short- and long-chain fatty acids, and monoglycerides (fats composed of one molecule of a fatty acid and one molecule of glycerol); these substances are transported to the liver via the portal venous system
   a. Lipase hydrolyzes the bonds between glycerol and fatty acids
   b. This process restores water molecules that were released when the bonds were formed
3. Glycerol diffuses directly through the mucosa
4. Short-chain fatty acids diffuse into the intestinal epithelial cells and are transported to the liver via the portal venous system
5. Long-chain fatty acids and monoglycerides in the intestine dissolve in the bile salt micelles
   a. They diffuse from the micelles into the intestinal epithelial cells
   b. In the endothelial cells, lipase breaks down absorbed monoglycerides into glycerol and fatty acids
6. Fatty acids and glycerol are recombined to form fats in the smooth endoplasmic reticulum of the epithelial cells
7. Triglycerides (fats), along with a small amount of cholesterol and phospholipid, are coated with a thin layer of protein to form lipoprotein particles called *chylomicrons*
   a. Chylomicrons collect in the intestinal lacteals (lymphatic vessels) and are transported through lymphatic channels
   b. Then chylomicrons enter the circulation through the thoracic duct and are distributed to body cells
      (1) In the cells, fats are extracted from the chylomicrons and broken down into fatty acids and glycerol by enzymes
      (2) They are absorbed and recombined in fat cells to reform tryglycerides (fat) for storage and later use

IV. Carbohydrate Metabolism

A. General information

1. *Metabolism* is the transformation of substances into energy or materials the body can use or store; it consists of two processes
   a. ***Anabolism*** is the synthesis of simple substances into complex ones
   b. ***Catabolism*** is the breakdown of complex substances into simpler ones or into energy
2. All ingested carbohydrates are converted to glucose, the body's main energy source
3. Glucose that is not needed for immediate energy requirements is stored as glycogen or converted to lipids

4. Energy from glucose catabolism is generated in three phases: ***glycolysis,*** the ***citric acid cycle*** (also called the Krebs, or tricarboxylic acid, cycle), and the ***electron transport system***
   a. Glycolysis, which occurs in the cell cytoplasm, does not require oxygen
   b. The citric acid cycle and electron transport system, which occur in mitochondria, require oxygen

B. Glycolysis

1. During glycolysis, enzymes break down the six-carbon glucose molecule into two three-carbon pyruvic acid (pyruvate) molecules; this process yields energy in the form of adenosine triphosphate (ATP)
2. If the oxygen supply to the tissues is inadequate, pyruvic acid is reduced by cytoplasmic enzymes to lactic acid by the addition of two hydrogen atoms
3. When adequate oxygen becomes available, lactic acid is oxidized back to pyruvic acid

C. Citric acid cycle

1. The citric acid cycle is a pathway by which a molecule of acetyl-coenzyme A (acetyl-CoA) is oxidized enzymatically to yield energy
2. In this phase of carbohydrate metabolism, pyruvic acid releases a CO_2 molecule and is converted in the mitochondria to a two-carbon acetyl fragment, which combines with coenzyme A (a complex organic compound) to form acetyl-CoA
3. Then the two-carbon acetyl fragments of acetyl-CoA enter the citric acid cycle by combining with the four-carbon compound oxaloacetic acid to form citric acid, a six-carbon compound; in this process, the coenzyme A molecule is detached from the acetyl group and becomes available to form more acetyl-CoA molecules
4. Enzymes then convert citric acid into intermediate compounds and eventually convert it back into oxaloacetic acid, which is available to repeat the cycle
5. Each turn of the citric acid cycle releases hydrogen atoms, which are picked up by the coenzymes nicotinamide adenine dinucleotide (NAD) and flavin adenine dinucleaotide (FAD); it also liberates carbon dioxide and generates energy

D. Electron transport system

1. In the electron transport system, carrier molecules on the inner mitochondrial membrane pick up electrons from the hydrogen atoms carried by NAD and FAD (each hydrogen atom consists of a hydrogen ion and an electron)
2. The carrier molecules transport the electrons through a series of enzyme-catalyzed oxidation-reduction reactions in the mitochondria
   a. During ***oxidation,*** a chemical compound loses electrons
   b. During ***reduction,*** a compound gains electrons
3. These reactions release the energy contained in the electrons and generate ATP
4. After passing through the electron transport system, the hydrogen ions produced in the citric acid cycle combine with oxygen to form water

E. Role of the liver and muscle cells

1. The liver plays an essential role in regulating blood glucose levels
   a. When glucose is present in quantities that exceed immediate needs, hormones stimulate the liver to convert glucose into glycogen or lipids
      (1) Glycogen is formed through ***glycogenesis***
      (2) Lipids are formed through ***lipogenesis***
   b. When the blood glucose level is inadequate, the liver can form glucose by two processes
      (1) The liver can break down glycogen to glucose through ***glycogenolysis***
      (2) It also can synthesize glucose from amino acids through ***gluconeogenesis***
2. Muscle cells can convert glucose to glycogen for storage, but they lack enzymes to convert glycogen back to glucose when needed
   a. During vigorous muscular activity, when oxygen demand exceeds the supply, muscle cells break down glycogen to yield lactic acid and energy
      (1) Lactic acid accumulates in the muscles, and muscle glycogen is depleted
      (2) Some lactic acid diffuses from muscle cells; it is transported to the liver and reconverted to glycogen
      (3) Then the liver converts the newly formed glycogen to glucose, which is transported through the bloodstream to the muscles and reformed into glycogen
   b. When muscular exertion ceases, part of the accumulated lactic acid is reconverted to pyruvic acid and then oxidized completely to yield energy by means of the citric acid cycle and electron transport system

V. Protein Metabolism

A. General information

1. Proteins are absorbed as amino acids; they are transported by the portal venous system to the liver and then throughout the body by blood
2. Absorbed amino acids mix with other amino acids in the body's amino acid pool; these other amino acids may be produced by protein breakdown or synthesized in the body from other substances, such as ketoacids
3. Amino acids cannot be stored; they are converted to protein or glucose or are catabolized to provide energy
4. For these changes to occur, amino acids must be transformed by ***deamination*** or ***transamination***
   a. In deamination, an amino group ($-NH_2$) is removed and the amino residue is excreted as urea
   b. In transamination, an amino group is exchanged for a keto group in a keto acid, through the action of transaminase enzymes; the process converts the amino acid to a keto acid and the original keto acid to an amino acid

B. Amino acid synthesis

1. Body proteins are synthesized from 20 different amino acids selected from the body's amino acid pool
2. The body can synthesize 12 amino acids from carbohydrates, fats, or other amino acids; they are called *nonessential amino acids*

3. The body cannot synthesize the other eight amino acids and must obtain them through dietary intake; they are called *essential amino acids*

C. Amino acid conversion

1. Amino acids that are not used for protein synthesis can be converted to keto acids and metabolized by the citric acid cycle and electron transport system to yield energy
   a. Some keto acids can enter the citric acid cycle directly by combining with one of the intermediate compounds in the cycle
   b. Others must undergo one of two possible conversions
      (1) They may be converted to pyruvic acid and then to acetyl-CoA, which combines with oxaloacetic acid to from citric acid
      (2) They may be converted directly to acetyl-CoA, which combines with oxaloacetic acid to form citric acid
2. Amino acids can be converted to other nutrients
   a. They can be converted to fats
      (1) Amino acids that are not used for protein synthesis may be converted to pyruvic acid and then to acetyl-CoA
      (2) The acetyl-CoA fragments condense to form long-chain fatty acids; this process is the reverse of fatty acid breakdown
      (3) These fatty acids combine with glycerol to form fats
   b. Amino acids also can be converted to glucose
      (1) They are converted to pyruvic acid
      (2) Pyruvic acid then may be converted to glucose

VI. Lipid Metabolism

A. General information

1. Fats are stored in adipose tissue within cells until required for energy
2. When required for energy, each fat molecule is hydrolyzed to glycerol and three molecules of fatty acids
3. Glycerol is converted to pyruvic acid and then to acetyl-CoA, which enters the citric acid cycle to yield energy
4. Long-chain fatty acids are catabolized into two-carbon fragments, which combine with coenzyme A to form acetyl-CoA fragments; the acetyl-CoA then enters the citric acid cycle to yield energy

B. Ketone body formation

1. The liver normally forms ketone bodies from acetyl-CoA fragments, which are derived primarily from fatty acid catabolism
2. Acetyl-CoA molecules yield three types of ketone bodies: acetoacetic acid, beta-hydroxybutyric acid, and acetone
   a. Acetoacetic acid forms when two acetyl-CoA molecules combine and coenzyme A is released from them

b. Beta-hydroxybutyric acid forms when hydrogen is added to the oxygen atom in the acetoacetic acid molecule; the designation beta refers to the location of the carbon atom that contains the hydroxyl (OH) group
c. Acetone forms when the carboxyl (COOH) group of acetoacetic acid releases CO_2

3. Muscle, brain, and other tissues oxidize these ketone bodies for energy
4. Certain conditions may cause production of more ketone bodies than the body can oxidize for energy
   a. Such conditions include fasting, starvation, and uncontrolled diabetes (in which the body cannot break down glucose)
   b. In all these conditions, the body must use fat, rather than glucose, as a primary energy source
   c. The resulting excess of ketone bodies disturbs the body's normal acid-base balance and homeostatic mechanisms, leading to *ketosis*

C. Lipid formation from proteins and carbohydrates

1. Excess amino acids can be converted to fat through keto acid–acetyl-CoA conversion
2. Glucose may be converted to pyruvic acid and then to acetyl-CoA, which is converted into fatty acids and then fat in the same way that amino acids are converted into fat

VII. Hormonal Regulation of Metabolism

A. General information

1. Blood glucose levels must remain within a certain range for the body to maintain its normal functions
2. Various hormones are secreted in response to changes in blood glucose level
3. These hormones stimulate metabolic processes that return the blood glucose level to normal

B. Hormones that increase blood glucose level

1. Glucagon promotes glycogen breakdown to glucose (glycogenolysis), amino acid conversion to glucose (gluconeogenesis), and lipid breakdown (lipolysis), which liberates free fatty acids and glycerol that can be converted to glucose
2. Epinephrine promotes glycogenolysis, gluconeogenesis, and lipolysis
3. Growth hormone (GH) has multiple effects
   a. It promotes protein synthesis by facilitating amino acid entry into cells
   b. It causes fat lipolysis from adipose tissue and promotes the use of fat rather than carbohydrate as an energy source
   c. It suppresses carbohydrate use for energy, causing blood glucose to rise as a result of reduced glucose use
   d. It promotes conversion of liver glucogen to glucose, which also tends to increase blood glucose level

4. Cortisol promotes protein hydrolysis to amino acids, which can be converted to glucose through gluconeogenesis
5. Thyroxine usually raises the blood glucose level by promoting gluconeogenesis and lipolysis

C. Hormones that decrease blood glucose level

1. Insulin is the only hormone that substantially lowers the blood glucose level
2. Insulin promotes cell uptake and use of glucose as an energy source
3. It promotes glucose storage as glycogen (glycogenesis) and lipids (lipogenesis)

Study Activities

1. Identify the major nutrients, including vitamins and minerals, and explain their chief functions.
2. Identify the major GI tract structures and their roles in digestion and nutrient absorption.
3. Describe the function of accessory organs in digestion and nutrient absorption.
4. Compare the functions of the major digestive secretions.
5. Contrast the digestion and absorption of carbohydrates, protein, and lipids.
6. Explain how glycolysis, the citric acid cycle, and the electron transport system cause carbohydrate metabolism.
7. Trace the processes of amino acid synthesis and conversion.
8. Discuss lipid metabolism and formation from proteins and carbohydrates.
9. Describe how hormones increase and decrease blood glucose levels.

10

Urinary System

Objectives

After studying this chapter, the reader should be able to:
- Identify the major urinary system structures and describe their functions.
- Describe the functions of the major parts of the nephrons.
- Discuss the role of glomerular filtration in urine production.
- Explain how hormones regulate urine volume and concentration.
- Describe how the countercurrent mechanism allows urine concentration.
- Explain how the kidneys regulate blood volume and blood pressure through the renin-angiotensin-aldosterone mechanism.
- Define the voiding reflex and explain how urination is under voluntary control.

I. Urinary System Structures

A. General information

1. The urinary system is the main excretory system of the body; it consists of two major structures
   a. The *kidneys* produce urine
   b. The *excretory duct system,* which consists of the calyces, pelvis, and ureters, transports urine to the bladder, which excretes urine through the urethra
2. The kidneys also perform other functions
   a. They help regulate water and electrolyte balance (see Chapter 11, Water, Electrolyte, and Acid-Base Balance, for more information)
   b. The kidneys also have endocrine functions
      (1) They produce the hormone *erythropoietin,* which regulates erythrocyte production in the bone marrow (see Chapter 7, Blood, for more information)
      (2) They also produce the enzyme *renin,* which helps regulate blood pressure

B. Kidneys

1. The kidneys are paired, bean-shaped organs that lie retroperitoneally in the lumbar region on each side of the vertebrae
2. Blood vessels and nerves enter and exit each kidney at the hilus, an indentation in the medial border
3. The kidneys have two regions: the cortex and medulla

a. The cortex is the outer layer of the kidney; portions of the cortex, called renal columns, extend deeper into the kidney
b. The medulla is the inner part of the kidney; it consists of up to 18 triangular renal pyramids separated by renal columns
   (1) The pyramid bases lie just below the cortex; the pyramid tips, or papillae, project into the renal calices
   (2) Urine drains from tubules at the tips of the papillae into the calices

C. Excretory duct system

1. The excretory duct system collects and excretes urine; it begins at the *renal pelvis* (expanded upper end of the ureters) and includes *ureters* (one from each kidney), *urinary bladder,* and *urethra*
2. The renal pelvis collects urine
   (1) Funnel-shaped sinus chambers, called minor calices, surround the renal pyramids
   (2) Minor calices join to form major calices
   (3) Major calices join to form the renal pelvis, the expanded upper end of the ureter
3. Each ureter is a muscular tube that leaves the kidney at the hilus; it transports urine from the kidneys to the bladder
4. The bladder is a hollow, spherical organ that acts as the body's urine reservoir; it is connected to the urethra
5. The urethra is a muscular tube that expels urine from the body; its exterior opening is called the urethral meatus

II. Nephrons

A. General information

1. *Nephrons* are the functional units of the kidneys
2. Each nephron consists of a *glomerulus* (located inside Bowman's capsule), *renal tubule,* and *juxtaglomerular apparatus*
3. Nephrons perform three basic functions
   a. They mechanically filter fluids, wastes, electrolytes, acids, and bases
   b. They reabsorb certain molecules
   c. They secrete certain molecules

B. Glomerulus

1. The glomerulus filters water and soluble material from the blood into the renal tubule
2. It consists of a capillary network in Bowman's capsule (the expanded proximal end of the renal tubule), located near the renal cortex
3. The glomerular capillaries have three layers: an inner fenestrated endothelium, a middle basement membrane, and an outer layer of specialized epithelial cells called podocytes, which are derived from the epithelium of Bowman's capsule
4. The glomerular capillary network receives blood from the afferent arteriole
5. Glomerular capillaries join and exit the glomerulus as the efferent arteriole; this arteriole separates into the vasa recta that surround the renal tubule

6. In the glomerulus, the capillary network is held together by groups of mesangial cells (special connective tissue cells) in the vascular pole; this is where the afferent and efferent arterioles enter and exit the glomerulus

C. Renal tubule

1. The renal tubule transports filtrate from the glomerulus to the renal calices
2. It is divided into three portions: proximal convoluted tubule, Henle's loop, and distal convoluted tubule
   a. The proximal convoluted tubule connects Bowman's capsule to Henle's loop
   b. The descending limb of Henle's loop dips into the renal medulla; the ascending limb returns to the renal cortex
   c. The distal convoluted tubule connects Henle's loop to a collecting tubule, which joins with other collecting tubules and empties into a calix to remove urine from the kidney

D. Juxtaglomerular apparatus

1. The juxtaglomerular apparatus regulates blood flow through the glomerulus and regulates blood pressure by producing renin
2. This apparatus consists of juxtaglomerular cells and the macula densa
   a. Juxtaglomerular cells are specialized cells that contain renin granules; they are located in the wall of the afferent arteriole, where the tubule is in contact with the arteriole
   b. The macula densa is an area of compact, heavily nucleated cells in the distal convoluting tubule, where the tubule makes contact with the vascular pole of the glomerulus

III. Urine Production

A. General information

1. The kidneys receive and filter a large volume of blood from the renal artery; tubular reabsorption and secretion converts glomerular filtrate into urine
2. Within the nephrons, glomeruli filter the blood; then the filtrate flows through the renal tubule
3. The tubules reabsorb and secrete various substances from the filtrate, changing its composition and concentration and ultimately producing urine
4. The glomerular filtration rate depends on three factors: glomerular capillary permeability, blood pressure, and effective filtration rate
5. The juxtaglomerular apparatus regulates glomerular filtration pressure by varying the glomerular filtration volume

B. Glomerular filtration

1. The glomeruli filter about 125 ml of water and dissolved materials every minute from the blood as it flows through their capillary network
2. The glomerular filtration rate depends on three factors: permeability of the glomerular capillary walls, blood pressure, and effective filtration rate
   a. A change in any of these factors can alter the glomerular filtration rate significantly
   b. For example, decreased cardiac output and blood pressure can reduce renal blood flow and the glomerular filtration rate

3. The juxtaglomerular apparatus helps maintain a stable glomerular filtration rate, despite wide variations in systemic arterial pressure
   a. A change in the glomerular filtration rate causes a change in filtration volume and sodium concentration of the filtrate
   b. The macula densa detects this change as the fluid flows into the distal convoluting tubule; it conveys the information to the juxtaglomerular cells in the walls of the adjacent afferent arterioles
   c. In response, the juxtaglomerular cells constrict or dilate the afferent arterioles and possibly the efferent arterioles to maintain a stable glomerular filtration pressure
   d. The changes the glomerular filtration pressure (regulated by the juxtaglomerular apparatus) adjust the filtration rate of each nephron based on the character (volume and sodium concentration) of the glomerular filtrate

C. Tubular reabsorption and secretion

1. As the filtrate passes through, the renal tubule selectively reabsorbs and secretes substances required by the body; reabsorption and secretion between tubular filtrate and peritubular blood occur via active and passive transport mechanisms
   a. Such substances as sodium, potassium, glucose, calcium, phosphates, and amino acids undergo active transport, which requires energy
   b. Such substances as urea, water, chloride, some bicarbonates, and some phosphates undergo passive transport, which does not require energy
2. Waste products and other unwanted substances are reabsorbed incompletely or not at all
3. Most reabsorption occurs in the proximal convoluted tubule, which reabsorbs such substances as water, glucose, amino acids, sodium ions, chloride ions, and other electrolytes; the tubule also secretes hydrogen ions, foreign substances, and creatinine
4. The proximal tubule also reabsorbs small amounts of protein that are filtered through the glomerular capillaries; the protein molecules are engulfed via pinocytosis, broken down by intracellular enzymes into amino acids, and reabsorbed
5. Henle's loop primarily reabsorbs sodium and chloride ions; it secretes sodium chloride
6. The distal convoluted tubule reabsorbs sodium, chloride, and bicarbonate ions as well as water; it secretes such substances as hydrogen (H^+) and ammonium (NH_3) ions, which help maintain the normal hydrogen ion concentration (pH) of body fluids (see Chapter 11, Water, Electrolyte, and Acid-Base Balance, for more information)
7. The distal tubule also secretes potassium (K^+) ions in exchange for sodium (Na^+) ions, which are reabsorbed from the filtrate; this exchange is controlled by the adrenal cortical hormone *aldosterone*
   a. Potassium and hydrogen ions compete for secretion
   b. Increased hydrogen ion secretion reduces potassium excretion; decreased hydrogen ion secretion increases potassium excretion
8. The collecting tubule can reabsorb or secrete sodium, potassium, hydrogen, and ammonium ions depending on the body's requirements; it also reabsorbs some water in the filtrate and secretes urea

9. Because of reabsorption in the renal tubule and collecting tubule, only about 1% of the original filtrate volume is excreted as urine

IV. Hormonal Regulation of Urine Volume and Concentration

A. General information

1. The hormones aldosterone and *antidiuretic hormone* (ADH) regulate urine volume and concentration
2. Aldosterone, a steroid hormone produced by the adrenal cortex, regulates the rate of sodium reabsorption from the tubules
3. ADH, a posterior pituitary hormone, is released in response to the solute concentration of blood flowing through the hypothalamus; it regulates water reabsorption from the collecting tubules

B. Aldosterone

1. This hormone acts primarily in the ascending limb of Henle's loop and the distal tubule
2. It promotes sodium reabsorption and potassium secretion
3. Because chloride ions are absorbed along with sodium, aldosterone indirectly promotes chloride reabsorption
4. Aldosterone secretion depends primarily on the sodium and potassium concentration in body fluids
   a. A decreased sodium concentration in body fluids stimulates aldosterone secretion; this increases sodium reabsorption, raising the sodium concentration in body fluids
   b. An increased sodium concentration in body fluids reduces aldosterone secretion; this reduces sodium reabsorption, decreasing the sodium concentration in body fluids
   c. An increased potassium concentration in body fluids also stimulates aldosterone secretion; this increases sodium reabsorption and potassium secretion by the tubules, raising the sodium concentration and lowering the potassium concentration in body fluids
   d. A decreased potassium concentration in body fluids inhibits aldosterone release; this reduces potassium secretion, raising the potassium concentration in body fluids

C. ADH

1. ADH regulates water reabsorption from the collecting tubules
   a. The hormone increases collecting tubule permeability, which allows them to absorb more water from the filtrate; this makes the urine more concentrated
   b. Without ADH, the collecting tubules are relatively impermeable to water; they reabsorb less water from the filtrate, and the urine remains dilute
2. Specialized cells in the hypothalamus, called *osmoreceptors,* regulate ADH release from the pituitary; these cells respond to osmotic pressure changes in the blood and body fluids
3. When body fluids are too concentrated (relatively low water content and relatively high solute content), osmotic pressure rises
   a. The pressure change stimulates osmoreceptors, which cause ADH secretion

b. ADH causes more water to be reabsorbed from the tubules, diluting body fluids and lowering the osmotic pressure

4. When body fluids are too dilute (relatively high water content and relatively low solute content), osmotic pressure decreases
   a. ADH is not released, and water is not reabsorbed from the tubules
   b. The excess water is excreted in dilute urine
   c. Water excretion increases the relative concentration of substances dissolved in the blood, raising the osmotic pressure toward normal

V. Countercurrent Mechanism

A. General information

1. The kidneys also can adjust body fluid concentrations
   a. When body fluids are too dilute, the kidneys eliminate excess water; this makes the fluids more concentrated
   b. When body fluids are too concentrated, the kidneys conserve water by excreting more concentrated urine; this makes the fluids more dilute
   c. The ***countercurrent mechanism*** allows the kidneys to concentrate urine
2. Body fluid concentration is expressed in terms of fluid ***osmolarity,*** which is a measure of the osmotic pressure exerted by substances dissolved in the fluid
   a. Upon leaving the glomerulus, filtrate normally has the same osmolarity as body fluids—about 300 milliosmols (mOsm) per liter; as filtrate passes along the renal tubule, its osmolarity changes in different parts of the nephron from 1,200 mOsm/liter in the descending limb of Henle's loop to 70 mOsm/liter in the collecting tubule
   b. The osmolarity of interstitial fluids in the kidney varies from 300 mOsm/liter in the cortex to 1,200 mOsm/liter in the medulla near the tips of the renal papillae
3. The ability of the kidneys to excrete urine of variable osmolarity depends on the anatomic arrangement of Henle's loop, *vasa recta* (peritubular capillaries that form long loops and have sluggish blood flow), and the collecting tubules, which pass through the medulla to empty at the tips of the renal papillae
   a. These three structures are adjacent to each other in the medulla
   b. They are surrounded by interstitial fluid
4. The mechanism by which urine is concentrated is called the countercurrent mechanism because filtrate passing through the limbs of Henle's loop flows in the opposite (countercurrent) direction from blood in the limbs of the vasa recta
   a. This mechanism is based on the theory of the countercurrent multiplication system
      (1) The system assumes the presence of two side-by-side tubes, each with a current flowing in opposite directions
      (2) It also assumes that the two tubes are joined at one end in a U-shape
      (3) Because tubular material is transported osmotically from one tube to another across the membrane separating them, the concentraton of material in the U-joint is greater than that entering or leaving the tubes
   b. This theory describes what occurs in the nephron
      (1) When the nephron is functioning normally, sodium is actively transported out of the solution in the proximal convoluted tubule, with water flowing passively

(2) The solution that enters the descending limb is *isotonic* (able to bathe cells without extracting water); at this point, however, sodium from extracellular fluid enters passively, so that when the solution in the limb reaches Henle's loop, it is highly concentrated and *hypertonic* (able to extract water from cells)
(3) When the solution enters the ascending limb, it flows along membranes that are impermeable to water but allow active transport of chlorine and passive flow of sodium
(4) When the solution enters the distal convoluted tubutle, it is *hypotonic,* which allows additional sodium and chlorine to be pumped out actively, with water flowing passively
(5) When the solution enters the collecting tubule, the action of ADH makes the membrane permeable to water and urea; this leaves a hypertonic solution that enters the renal pelvis as urine

B. Countercurrent mechanism function

1. Glomerular filtrate flows through the proximal tubule, which reabsorbs water and dissolved substances in equal proportions; this reabsorption reduces the filtrate volume by 80% but does not change its osmolarity
2. The filtrate flows into the descending limb of Henle's loop, which is freely permeable to water and sodium chloride
3. As filtrate passes down the descending limb, it is exposed to the high osmolarity of the interstitial fluid in the medulla
   a. Water moves from the descending limb into the interstitial tissue by osmosis
   b. Sodium chloride diffuses into the tubular filtrate from the interstitial tissue, where the sodium chloride concentration is much higher
   c. As a result, tubular filtrate osmolarity increases as the fluid passes down the descending limb until its osmolarity is as high as that of the interstitial fluid in the medulla
4. The filtrate moves into the ascending limb of Henle's loop
   a. This limb is relatively impermeable to water and possesses an aldosterone-regulated transport mechanism by which sodium is pumped out of the filtrate
   b. Chloride moves out passively with the sodium, but water remains in the filtrate
   c. As a result, tubular filtrate osmolarity falls as the filtrate moves through the ascending limb
5. When the filtrate reaches the distal tubule, its osmolarity has fallen to 100 mOsm/liter; it continues to fall to as low as 70 mOsm/liter until it enters the collecting tubule
6. ADH regulates the final urine concentration
   a. If the body has insufficient water, ADH is secreted and collecting tubule permeability increases
   b. This allows water to move out of the collecting tubule by osmosis into the hypertonic interstitial fluid, which produces concentrated urine for excretion
   c. If the body has excess water, ADH is not excreted; the collecting tubules remain impermeable to water, which produces a dilute urine for excretion

VI. Renal Regulation of Blood Pressure and Blood Volume

A. General information

1. Juxtaglomerular cells in the kidneys regulate blood pressure and blood volume by secreting renin
2. They secrete renin in response to decreased blood pressure, blood volume, or plasma sodium concentration

B. Regulation process

1. Renin interacts with a blood protein called renin substrate to yield the polypeptide angiotensin I
2. Angiotensin I is converted to angiotensin II by enzymes in endothelial cells present in the lung and other tissues
3. Angiotensin II raises blood pressure by increasing vasoconstriction and stimulating the adrenal cortex to secrete aldosterone
4. Aldosterone increases sodium and water reabsorption in the renal tubule, which increases blood volume
5. The renin-angiotensin-aldosterone system is self-regulating
   a. Blood pressure, blood volume, and sodium concentration rise in response to renin secretion
   b. When these levels reach the normal range, the juxtaglomerular cells are no longer stimulated, causing renin secretion to fall (see *Kidney Regulation of Blood Pressure and Blood Volume* for an illustration of the process)

VII. Urine Elimination

A. General information

1. After urine is formed in the nephrons, it passes from the collecting tubules through the minor and major calyces of the kidney to the renal pelvis
2. From the pelvis, urine is conveyed by the ureters to the bladder
   a. Peristaltic contractions of the smooth muscle wall move the urine down the ureter
   b. Peristaltic contractions occur at a rate of about 1 to 5 per minute
3. The bladder stores urine until the ***voiding reflex*** is triggered
   a. The bladder can hold 500 to 600 ml in an adult
   b. Accumulation of 300 to 400 ml usually triggers the *voiding reflex*
4. Urine elimination results from involuntary (reflex) and voluntary (learned and intentional) processes
5. Urine flows from the bladder through the urethra; it is expelled from the body through the external urethral opening

B. Voiding reflex

1. The voiding reflex typically is activated when the bladder contains 300 to 400 ml of urine
2. As urine fills the bladder, it stretches the bladder walls; this triggers the voiding reflex
3. Stretch receptors in the bladder walls transmit sensory impulses to the spinal cord, which stimulates parasympathetic neurons

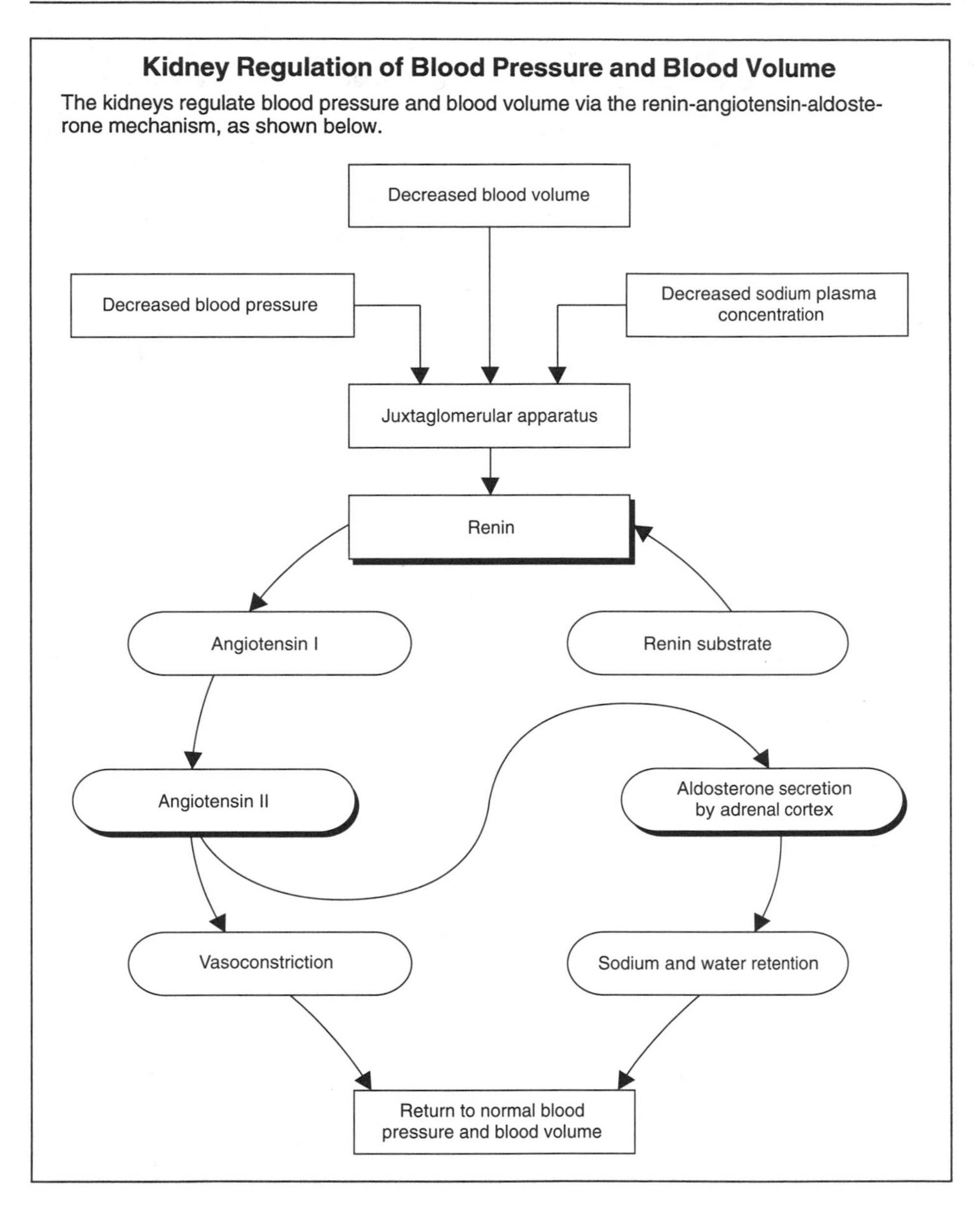

4. The spine relays motor impulses that cause bladder wall contraction and relaxation of the external urethral sphincter; this leads to urination unless voluntary control is exerted
5. Besides being transmitted to the spinal cord, sensory impulses from the bladder walls are sent to higher brain centers; they are interpreted as a sense of bladder fullness or the need to urinate
6. The brain can inhibit or stimulate the voiding reflex

a. When urination must be delayed, the individual inhibits the reflex by contracting the external sphincter, which is composed of striated muscle and is under voluntary control
b. When the opportunity to urinate becomes available, the individual voluntarily contracts the abdominal muscles, which raises intra-abdominal pressure and assists in expelling urine from the bladder

Study Activities

1. Describe the functions of the urinary system structures.
2. Trace the flow of urine through the structures of the kidneys and the excretory duct system.
3. Identify the parts of the nephron and explain their functions.
4. Compare the effects of glomerular filtration and tubular reabsorption and secretion on urine production.
5. Explain how aldosterone and ADH regulate urine volume and concentration.
6. Discuss the principles by which the countercurrent mechanism functions.
7. Describe the process by which the kidneys regulate blood pressure and blood volume.
8. Contrast the voluntary and involuntary aspects of urine elimination.

11

Water, Electrolyte, and Acid-Base Balance

Objectives

After studying this chapter, the reader should be able to:
- Identify the two major body fluid compartments and their contents.
- Identify the routes by which water enters and leaves the body, and explain the mechanisms of water balance.
- Differentiate among the major cations and anions.
- Compare the mechanisms of electrolyte balance.
- Define pH and describe its importance in acid-base balance.
- Identify common sources of hydrogen ions.
- Explain how the buffer systems help maintain the normal pH of body fluids.
- Compare the roles of the lungs and kidneys in maintaining acid-base balance.
- Describe the factors that affect bicarbonate formation in the renal tubule epithelial cells.

I. Body Fluids

A. General information

1. Homeostasis depends on a complex interrelationship among water, electrolyte, and acid-base metabolism
2. More than 50% of the average adult's body consists of water
3. The proportion of body water varies inversely with the body's fat content because fat contains no water
    a. An obese individual has a lower percentage of water than a lean person
    b. Most women have a lower percentage of water than men because their bodies normally have a higher percentage of body fat
4. Body water contains dissolved substances (solutes) that are necessary for physiologic functioning; solutes include ***electrolytes,*** glucose, amino acids, and other nutrients
5. Body fluid composition differs by compartment

B. Body fluid compartments

1. The ***intracellular fluid compartment*** consists of the fluid within the body's cells
2. The *intravascular fluid compartment* consists of the fluid in blood plasma and lymphatic system
3. The *interstitial fluid compartment* consists of fluid distributed diffusely through the loose tissue surrounding the cells

4. Intravascular and interstitial fluids are separated by a capillary endothelium that is freely permeable to water, electrolytes, and other solutes
   a. Therefore, intravascular and interstitial fluids are similar in composition
   b. The intravascular and interstitial fluid compartments commonly are grouped together in a single compartment called the ***extracellular fluid compartment***
5. The composition of intracellular fluid differs from that of extracellular fluid
   a. Intracellular fluid has higher concentrations of protein, potassium, magnesium, phosphate, and sulfate
   b. Intracellular fluid has lower concentrations of sodium, calcium, chloride, and bicarbonate
6. Active transport helps maintain different concentrations of sodium and potassium in intracellular and extracellular fluids (see Chapter 1, Human Cell, for more information)

C. Body fluid osmolarity

1. When a semipermeable membrane separates two solutions of unequal solute concentration, water shifts by *osmosis* from the less concentrated solution to the more concentrated solution
2. The ability of the more concentrated solution to attract water is called its *osmotic activity*
3. The solution's osmotic activity depends on the number of particles dissolved in the solution
   a. Osmotic activity is unrelated to the particles' molecular weight or valence
   b. The same osmotic activity results from a solution that contains equal numbers of sodium ions (which are monovalent), calcium ions (which are divalent), or glucose molecules (which do not dissociate into ions)
   c. The *osmotic pressure* of a solution (pressure exerted by a solution on a semipermeable membrane) usually is measured in terms of *osmolarity*
      (1) A solution of 1 liter of water that contains 1 gram molecular weight of a substance that does not dissociate in solution (such as glucose) has an osmolarity of 1 osmol per liter (1 Osm/liter)
      (2) A solution of 1 liter of water that contains 1 gram molecular weight of an electrolyte that dissociates into two ions (such as sodium chloride, NaCl) has an osmolarity of 2 Osm/liter
      (3) A solution of 1 liter of water that contains 1 gram molecular weight of an electrolyte that dissociates into three ions (such as calcium chloride, $CaCl_2$) has an osmolarity of 3 Osm/liter
   d. Because body fluids have low concentrations of dissolved particles, their osmolarity usually is expressed in milliosmols per liter (mOsm/liter)

II. Water Balance

A. General information

1. Water is essential to normal physiologic functioning
2. The body gains and loses water daily through fluid intake and output; water enters the body via the gastrointestinal (GI) tract and leaves via the skin, lungs, GI tract, and urinary tract

3. These gains and losses must be balanced to stabilize the body's water content and to permit proper physiologic functioning
4. Two mechanisms help maintain water balance: thirst, which regulates water intake, and the *countercurrent mechanism,* which regulates urine concentration

B. Water intake

1. Water normally enters the body from the GI tract
2. Each day, the body derives about 1,500 ml of water from consumed liquids
3. The body also receives 700 ml from consumption of solid foods, which may contain up to 97% water
4. Food oxidation in the body generates carbon dioxide and 250 ml of water (water of oxidation)

C. Water output

1. Water leaves the body through the skin (in perspiration), lungs (in exhaled air), GI tract (in feces), and urinary tract (in urine)
2. Each day, the body loses 800 ml of water through the skin and lungs; this amount of water loss may increase dramatically with strenuous exertion
3. Although the GI tract contents include large amounts of fluid, the colon normally absorbs almost all the water; the body loses only about 200 ml of water in feces
4. Urine excretion is the main route of water loss; output typically varies from 1,000 to 1,500 ml daily

D. Mechanisms of water balance

1. Two mechanisms help maintain water balance: thirst and the countercurrent mechanism
2. Thirst (conscious desire for water) primarily regulates water intake
   a. Dehydration decreases the extracellular fluid volume, which increases its sodium concentration and osmolarity
   b. When the sodium concentration reaches about 2 mEq/liter above normal, it stimulates the neurons of the thirst center in the hypothalamus
   c. When the brain directs motor neurons to satisfy thirst, the individual drinks the proper amount of water to restore extracellular fluids to normal
3. Through the countercurrent mechanism, the kidneys can regulate water output by excreting urine of greater or lesser concentration (see Chapter 10, Urinary System, for more information)

III. Electrolyte Balance

A. General information

1. *Electrolytes* are substances that dissociate into *ions* (electrically charged particles) when dissolved in water; normal metabolism and function require sufficient quantities of each major electrolyte and proper balance among electrolytes
2. Ions may be positively charged *cations* or negatively charged *anions*
3. Intracellular and extracellular fluids normally contain different concentrations of electrolytes
4. Electrolyte balance is maintained by various mechanisms

B. Electrolytes

1. Major cations include sodium (Na^+), potassium (K^+), calcium (Ca^{++}), and magnesium (Mg^{++})
2. Major anions include chloride (Cl^+), bicarbonate (HCO_3^+), and phosphate (HPO_4^+)
3. Normally, the electrical charges of the cations and anions are balanced so that body fluids are electrically neutral
4. Ion concentration is expressed in terms of the ion's *equivalent weight* (ability to combine with other ions); equivalent weight equals the ion's gram molecular weight (amount of a substance that has a weight in grams equal to its molecular weight) divided by its chemical valence (numerical expression of chemical combining capacity)
   a. Ions with the same number of equivalents in a solution have equal combining powers; their concentrations are considered equal even though their gram molecular weights are different
   b. In body fluids, ions are present in such low concentrations that they usually are expressed in milliequivalents per liter (mEq/liter)
5. Intracellular and extracellular fluids normally have different electrolyte compositions because their cells are permeable to different substances
   a. Sodium concentration in intracellular fluid is 10 mEq/liter; in extracellular fluid, 136 to 146 mEq/liter
   b. Potassium concentration in intracellular fluid is 140 mEq/liter; in extracellular fluid, 3.6 to 5 mEq/liter
   c. Calcium concentration in intracellular fluid is 10 mEq/liter; in extracellular fluid, 4.5 to 5.8 mEq/liter
   d. Magnesium concentration in intracellular fluid is 40 mEq/liter; in extracellular fluid, 1.6 to 2.2 mEq/liter
   e. Chloride concentration in intracellular fluid is 4 mEq/liter; in extracellular fluid, 96 to 106 mEq/liter
   f. Bicarbonate concentration in intracellular fluid is 10 mEq/liter; in extracellular fluid, 24 to 28 mEq/liter
   g. Phosphate concentration in intracellular fluid is 100 mEq/liter; in extracellular fluid, 1 to 1.5 mEq/liter

C. Mechanisms of electrolyte balance

1. Homeostasis depends on a complex interrelationship among water, electrolyte, and acid-base metabolism; electrolytes profoundly affect water distribution, osmolarity, and acid-base balance
2. The body uses various mechanisms to maintain electrolyte balance (see Chapter 10, Urinary System, for more information on selected mechanisms)
   a. Sodium is regulated chiefly by the kidneys through the action of aldosterone; it is absorbed readily from food by the small intestine and is excreted through the skin and kidneys
   b. Potassium also is regulated by the kidneys through the action of aldosterone; most of it is absorbed from food in the GI tract, and the amount of potassium excreted in urine normally equals dietary potassium intake
   c. Calcium in blood is in equilibrium with calcium salts in bone; calcium is regulated primarily by parathyroid hormone, which controls calcium uptake from the intestinal tract and calcium excretion by the kidneys

- d. Magnesium is regulated by aldosterone, which controls renal reabsorption of magnesium; it is absorbed from the GI tract and excreted in urine, milk, and saliva
- e. Chloride is regulated by the kidneys; chloride ions move with sodium ions
- f. Bicarbonate is regulated by the kidneys, which may excrete, absorb, or form bicarbonate; it plays an important role in regulating *acid-base balance*
- g. Phosphate is regulated by the kidneys; it is absorbed well from food, incorporated with calcium in bone, and regulates by parathyroid hormone along with calcium

IV. Acid-Base Balance

A. General information

1. Acid-base balance results in a stable hydrogen ion (H^+) concentration in body fluids
2. An acid is a substance that dissociates in water and releases hydrogen ions
    - a. A strong acid is one that dissociates virtually completely, releasing a large number of hydrogen ions
    - b. A weak acid does not dissociate readily and releases fewer hydrogen ions
3. A base is a substance that dissociates in water and releases ions that can combine with hydrogen ions, such as hydroxyl ions (OH^-)
    - a. A strong base is one that dissociates virtually completely, releasing a large number of ions
    - b. A weak base does not dissociate readily and releases fewer ions
4. The hydrogen ion concentration of a fluid determines whether it is acidic or basic (alkaline)
    - a. A neutral solution, such as pure water, dissociates only slightly; it contains 0.0000001 (one ten-millionth) gram molecular weight (mol) of hydrogen ions per liter and the same amount of hydroxyl ions
    - b. This minute quantity may be expressed in exponential form as 10^7 g/liter
    - c. More commonly, hydrogen ion concentration is expressed as ***pH,*** which is the value of the exponent without the minus sign; for example, a neutral solution that contains 10^7 mols of hydrogen ions per liter has a pH of 7
        - (1) An acidic solution contains more hydrogen ions; its pH is less than 7
        - (2) An alkaline solution contains fewer hydrogen ions; its pH is greater than 7
    - d. Because pH is an exponential expression, a change of 1 pH unit represents a tenfold change in hydrogen ion concentration; for example, a solution with a pH of 6 has ten times more hydrogen ions than one with a pH of 7

B. Sources of hydrogen ions

1. The body is an acid-producing organism
2. Protein catabolism produces several nonvolatile acids, such as sulfuric, phosphoric, and uric acid
3. Fat oxidation produces acid ketone bodies (acetoacetic acid and beta-hydroxybutyric acid)
4. Anaerobic glucose catabolism produces lactic acid

5. Intracellular metabolism creates a large quantity of carbon dioxide as a by-product; some of the carbon dioxide dissolves in body fluids to form carbonic acid

V. Mechanisms of Acid-Base Balance

A. General information

1. ***Buffer systems*** and the lungs and kidneys maintain the blood pH within a narrow range —7.38 to 7.42 —by neutralizing and eliminating acids as rapidly as they are formed; these actions help maintain acid-base balance
2. The *sodium bicarbonate – carbonic acid buffer system* is the principal buffer in the extracellular fluid
3. The lungs affect acid-base balance by excreting carbon dioxide and regulating the carbonic acid content of the blood
4. The kidneys regulate acid-base balance by allowing tubular filtrate reabsorption of bicarbonate and by forming bicarbonate

B. Buffer systems

1. Buffer systems minimize pH changes caused by excess acids or bases
2. A buffer system consists of a weak acid and a salt of that acid, or a weak base and its salt
3. The buffer system reduces the effect of a sudden change in hydrogen ion concentration by converting a strong acid or base, which normally would dissociate completely, into a weak acid or base, which releases a smaller number of free hydrogen or hydroxyl ions
4. The sodium bicarbonate – carbonic acid buffer system is the principal buffer; others include the phosphate system and protein system
5. The pH of any buffer system depends on the ratio of the two components in the buffer and not on their absolute amounts
6. The sodium bicarbonate – carbonic acid buffer system is the principal buffer in extracellular fluid
   a. This system works better in vivo than in vitro; if acid were added continuously to a fixed amount of bicarbonate – carbonic acid buffer in a beaker, eventually all of the sodium bicarbonate would be consumed by neutralizing the acid, and the buffer would lose its effectiveness because it no longer would contain bicarbonate
   b. In the body, this buffer system maintains its efficiency for two reasons
      (1) Both components of the buffer are replenished continually
      (2) The concentration of both components is regulated physiologically — sodium bicarbonate by the kidneys and carbonic acid by the lungs
   c. In the sodium bicarbonate – carbonic acid buffer system, the normal ratio of the components (20 parts sodium bicarbonate to 1 part carbonic acid) maintains a pH of 7.4
      (1) A change in the 20:1 ratio causes a corresponding change in the pH of the buffer and the body fluids it regulates
      (2) This relationship can be visualized in terms of a board on a fulcrum
         (a) Normally, one end of the board is weighted by 20 parts sodium bicarbonate, and the other end is weighted by 1 part carbonic acid.
         (b) The fulcrum is positioned so the board balances at a pH of 7.4

(c) Changes in the ratio of sodium bicarbonate to carbonic acid will unbalance the board and shift the system to a higher or lower pH

7. The *phosphate buffer system* also helps maintain normal pH, especially in extracellular fluids
   a. The phosphate buffer uses sodium monohydrogen phosphate ($NaHPO_4$) as the acidic component and sodium dihydrogen phosphate (NaH_2PO_4) as the alkaline component
   b. This buffer system is especially important in neutralizing hydrogen ions secreted by the renal tubules
8. The *protein buffer system* helps maintain normal pH; in this system, intracellular proteins functions as buffers by absorbing hydrogen ions generated by the body's metabolic processes

C. Lungs

1. The lungs excrete carbon dioxide (CO_2) and regulate the carbonic acid (H_2CO_3) content of the blood
   a. Carbonic acid is derived from the carbon dioxide and water (H_2O) released as a byproduct of cellular metabolic activity
   b. Carbon dioxide is soluble in blood plasma
      (1) Some of the dissolved gas reacts with water to form carbonic acid, a weak acid that partially dissociates to form hydrogen and bicarbonate ions
      (2) All three substances are in equilibrium:
      $$CO_2 + H_2O \leftrightarrow H_2CO_3 \leftrightarrow H^+ + HCO_3^-$$
   c. Carbon dioxide dissolved in plasma is in equilibrium with the carbon dioxide in the pulmonary alveoli
      (1) The alveolar carbon dioxide concentration is expressed as a partial pressure (PCO_2)
      (2) Consequently, an equilibrium exists between alveolar PCO_2 and the various forms of carbon dioxide in the plasma:
      $$PCO_2 \leftrightarrow CO_2 + H_2O \leftrightarrow H_2CO_3 \leftrightarrow H^+ + HCO_3^-$$
2. The carbon dioxide content of alveolar air and the alveolar PCO_2 vary with the rate and depth of respirations
3. A change in alveolar PCO_2 causes a corresponding change in the amount of carbonic acid formed by dissolved carbon dioxide; these changes stimulate the respiratory center to alter the respiratory rate and depth
   a. A rise in alveolar PCO_2 increases the blood concentration of carbon dioxide and carbonic acid, which stimulates the respiratory center to increase the rate and depth of respirations; this stimulation lowers alveolar PCO_2 and leads to a corresponding decrease in the carbonic acid and carbon dioxide concentrations in blood
   b. A decrease in the rate and depth of respirations has an opposite effect, elevating the alveolar PCO_2, which leads to a corresponding increase in the carbon dioxide and carbonic acid concentrations in blood

D. Kidneys

1. The kidneys excrete various acid waste products
2. They also regulate the bicarbonate concentration in the blood in two ways

a. They allow bicarbonate reabsorption from tubular filtrate
b. They can form additional bicarbonate to replace that used in buffering acids
3. Bicarbonate recovery and formation in the kidneys depend on hydrogen ion secretion by the renal tubules in exchange for sodium ions, which simultaneously are reabsorbed from the tubular filtrate into the circulation
4. The renal tubules secrete hydrogen ions
a. Under the influence of the enzyme carbonic anhydrase, tubular epithelial cells form carbonic acid from carbon dioxide and water
b. This carbonic acid rapidly dissociates into hydrogen and bicarbonate ions
c. The hydrogen ions enter the tubular filtrate in exchange for sodium ions
d. The bicarbonate ions enter the bloodstream along with sodium ions that have been absorbed from the filtrate
5. Bicarbonate is reabsorbed from tubular filtrate
a. Each hydrogen ion secreted into tubular filtrate (in exchange for a sodium ion) combines with a bicarbonate ion to form carbonic acid, which rapidly dissociates into carbon dioxide and water
b. The carbon dioxide diffuses into the tubular epithelial cell, where it can combine with more water to form more carbonic acid
c. The leftover water molecule in the tubular filtrate is excreted in urine
d. As each hydrogen ion moves into tubular filtrate to combine with a bicarbonate ion there, a bicarbonate ion in the tubular epithelial cell diffuses into the circulation
e. This process is called bicarbonate reabsorption, even though the bicarbonate ion that enters the circulation is not the same one as in the tubular filtrate
6. To form more bicarbonate, the kidneys must secrete additional hydrogen ions in exchange for sodium ions
a. The renal tubules cannot continue to secrete hydrogen ions unless the excess ions can be combined with other substances in the filtrate and excreted
b. Excess hydrogen ions in the filtrate may combine with ammonia, which is produced by the renal tubules, or with phosphate salts, which are present in tubular filtrate
(1) Ammonia (NH_3) is formed in the tubular epithelial cells by removal of the amino groups from glutamine (an amino acid derivative) and from amino acids that are delivered from the circulation to the tubular epithelial cells
(a) Ammonia diffuses into the filtrate and combines with the secreted hydrogen ions to form ammonium ions (NH_4^+), which are excreted in the urine with chloride and other anions
(b) Each ammonia molecule secreted eliminates one hydrogen ion in the filtrate
(c) Simultaneously, sodium ions that have been absorbed from the filtrate and exchanged for hydrogen ions enter the circulation; so does the bicarbonate formed in the tubular epithelial cells
(2) Some secreted hydrogen ions combine with Na_2HPO_4, a disodium phosphate salt in the tubular filtrate

(a) Each secreted hydrogen ion that combines with the disodium salt converts it to the monosodium salt NaH_2PO_4
(b) This reaction forms a sodium ion, which is absorbed into the circulation along with a newly formed bicarbonate ion

7. Two factors affect the rate of bicarbonate formation by the renal tubular epithelial cells: the amount of dissolved carbon dioxide in the plasma and the potassium content of the tubular cells
   a. If the amount of carbon dioxide in the plasma increases, the renal tubular cells form more bicarbonate
      (1) Increased plasma carbon dioxide promotes increased carbonic acid formation by the renal tubular cells
      (2) Carbonic acid partially dissociates, yielding more hydrogen ions for excretion into the tubular filtrate and additional bicarbonate ions for entry into the circulation
      (3) This raises the plasma bicarbonate level and decreases the plasma level of dissolved carbon dioxide toward normal
   b. If the amount of carbon dioxide in the plasma decreases, the renal tubular cells form less carbonic acid
      (1) Fewer hydrogen ions are formed and excreted
      (2) This causes fewer bicarbonate ions to enter the circulation
      (3) The plasma bicarbonate level falls correspondingly
   c. The potassium content of the renal tubular cells also regulates plasma bicarbonate concentration by influencing the rate at which the renal tubules secrete hydrogen ions
      (1) Tubular cell potassium content and hydrogen ion secretion are interrelated
      (2) The rates at which potassium and hydrogen ions are secreted vary inversely (see Chapter 10, Urinary System, for more information)
         (a) If tubular secretion of potassium ions falls, hydrogen ion secretion rises
         (b) If tubular secretion of potassium ions increases, hydrogen ion secretion declines
      (3) Each hydrogen ion secreted into the tubular filtrate is accompanied by the addition of a bicarbonate ion to the blood plasma
         (a) Therefore, the plasma bicarbonate content rises when tubular secretion of hydrogen ions increases
         (b) For example, if vomiting or diarrhea cause potassium depletion, then potassium secretion by the tubular epithelial cells falls, and hydrogen ion secretion rises
         (c) When body potassium is depleted, more bicarbonate enters the circulation, and the plasma bicarbonate level increases above normal
      (4) The tubules excrete more potassium when the body contains excess potassium; when this occurs, fewer hydrogen ions are secreted, less bicarbonate is formed, and the plasma bicarbonate concentration decreases

Study Activities

1. Describe the composition and osmolarity of the intracellular and extracellular fluid compartments.
2. Discuss normal water intake and output.
3. Explain how thirst and the countercurrent mechanism help the body achieve water balance.
4. Identify the concentration (in mEg/liter) of the major cations and anions in intracellular and extracellular fluids.
5. Describe the mechanisms of electrolyte balance for the major cations and anions.
6. Explain how hydrogen ions affect acid-base balance.
7. Discuss how changes in ion ratios can affect acid-base balance in the sodium bicarbo-ate – carbonic acid buffer system.
8. Explain how the lungs help maintain acid-base balance.
9. Trace the movement of hydrogen and bicarbonate ions into and out of the renal tubule epithelial cells.

12

Endocrine System

Objectives

After studying this chapter, the reader should be able to:

- Differentiate between the two major types of glands.
- Describe the general function of hormones and explain how feedback mechanisms control their levels.
- Discuss the roles of hormone receptors and cyclic adenosine monophosphate in target cell response to hormones.
- Explain how the hypothalamus controls pituitary hormone secretion.
- Compare the functions of the major pituitary hormones.
- Discuss the synthesis, secretion, and effects of thyroid hormone.
- Explain how parathyroid hormone, vitamin D, and calcitonin regulate blood calcium.
- Identify the major hormones produced by the adrenal cortex and medulla, and compare their effects.
- Discuss the functions of the hormones produced by the islets of Langerhans.

I. Endocrine Glands

A. General information

1. Two major types of glands exist: exocrine and endocrine glands
   a. *Exocrine glands* discharge secretions through a duct onto an epithelial surface; these secretions have a lubricating or digestive function
   b. *Endocrine glands* discharge secretions directly into the bloodstream; these secretions, called *hormones,* help regulate metabolic processes
2. The major endocrine glands are the *pituitary, thyroid, parathyroid,* and *adrenal glands; the islets of Langerhans;* and the *ovaries* and *testes* (see Chapter 13, Reproductive System, for more information on ovarian and testicular secretions)
3. Some organs with nonendocrine functions contain groups of specialized endocrine cells that also secrete hormones
   a. The kidneys secrete renin and erythropoietin (see Chapter 10, Urinary System, for more information)
   b. The gastrointestinal (GI) tract secretes gastrin and other hormones (see Chapter 9, Nutrition, Digestion, and Metabolism, for more information)
   c. The placenta also has endocrine functions (see Chapter 14, Fertilization, Pregnancy, Labor, and Lactation)
4. Hormone output may be controlled directly or indirectly by feedback mechanisms

a. Output can be regulated *directly* by the hormone level produced by a gland
b. Output can be regulated *indirectly* by the level of a substance under hormonal control, such as glucose or sodium

5. Usually, an elevated level of hormone or hormone-regulated substance suppresses further hormone output; this arrangement is called a ***negative feedback mechanism***
6. Less commonly, a rising hormone level stimulates hormone output; this arrangement is called a *positive feedback mechanism*

B. Hormone receptors

1. Specific hormone receptors within target cells or on their cell membranes determine the cell's responsiveness to a specific hormone
2. The number of hormone receptors in target cells varies with changes in circulating hormone levels
   a. When hormone levels are above normal, the number of receptors decreases; this compensates for the hormone excess by reducing target cell affinity for the hormone
   b. When hormone levels are below normal, the number of receptors increases; this increases target cell affinity for the hormone

C. Cyclic adenosine monophosphate and hormone function

1. Some hormones (steroid and thyroid hormones) can enter the cell and bind with intracellular hormone receptors to exert their effect
2. However, many hormones cannot enter the target cell and must bind to receptors on the cell membrane
3. Receptor binding to the cell membrane activates other enzymes to cause the desired effect in the cell
   a. Receptor binding activates the enzyme adenylate cyclase, located on the inner surface of the cell membrane
   b. This catalyzes the conversion of intracellular adenosine triphosphate (ATP) into *cyclic adenosine monophosphate (AMP);* this conversion continues as long as the hormone acts on the cell membrane
   c. Cyclic AMP activates certain intracellular enzymes in the target cells
   d. The activated enzymes perform a specific function based on target cell characteristic; for example, enzymes cause glucose liberation from glycogen in liver cells or lipid synthesis in fat cells
   e. Duration of action of cyclic AMP is brief because the intracellular enzyme phosphodiesterase degrades cyclic AMP and converts it into the inactive form, AMP
   f. Through this process, a hormone attached to the exterior of a cell can exert intracellular effects

II. Pituitary Gland

A. General information

1. The pituitary gland is a small, pea-shaped gland connected to the hypothalamus by a narrow stalk
2. The gland lies in a small depression at the base of the skull (sella turcica) just behind the optic chiasm and optic nerves

3. It has an anterior lobe and a posterior lobe, which are connected by a small rudimentary intermediate lobe, which has no function in humans
   a. Releasing factors from the hypothalamus regulate hormone release from anterior lobe cells
   b. After being synthesized in the hypothalamus, posterior lobe hormones are transported by nerve axons to the posterior lobe cells, where they are stored; then they are released from the posterior lobe in response to nerve impulses transmitted from the hypothalamus down the pituitary stalk

B. Hypothalamic control

1. The hypothalamus, a portion of the diencephalon of the brain, activates, controls, and integrates various endocrine functions
2. It produces releasing and inhibiting factors (regulatory hormones) that control the anterior pituitary; these factors enter the pituitary through portal venous pathways
3. The hypothalamus also produces hormones that are stored in and secreted by the posterior pituitary
4. A negative feedback mechanism controls the release of hypothalamic substances
   a. Cells in the hypothalamus regulate the level of most pituitary hormones; these cells continually monitor the levels of circulating hormones produced by target glands, such as the thyroid gland
   b. When the target gland hormone level declines, the hypothalamus produces releasing factors that are carried through the portal venous pathways to the pituitary; the factors induce the release of ***tropic hormones*** that stimulate hormone production by the target gland
   c. The level of the hormone produced by the target gland rises until it reaches the upper range of normal; then the elevated hormone level "shuts off" further release of releasing factors and tropic hormones
5. The mechanism maintains a relatively steady hormone output from the target gland and prevents wide fluctuations in hormone levels that might disrupt normal body functions
6. Higher cortical centers also can affect hypothalamic control of hormone release; for example, pituitary secretion may change in response to strong emotions, such as anxiety, anger, or fear (see *Mechanism of Hypothalamic Control,* page 138, for an illustration of the process)

C. Anterior lobe

1. The anterior lobe is composed of cords of epithelial cells that contain granules of stored hormone
2. Five different cell types secrete six hormones
   a. Somatotropes secrete growth hormone (GH), or somatotropin
   b. Mammotropes secrete prolactin
   c. Thyrotropes secrete thyroid-stimulating hormone (TSH), or thyrotropin
   d. Corticotropes secrete adrenocorticotropic hormone (ACTH), or corticotropin
   e. Gonadotropes secrete follicle-stimulating hormone (FSH) and luteinizing hormone (LH)
3. Anterior pituitary hormones perform various functions
   a. GH promotes tissue growth in different ways
      (1) It stimulates the liver to secrete small proteins called somatomedins, which are responsible for the cellular effects of the hormone

Mechanism of Hypothalamic Control

Through a feedback mechanism, the hypothalamus regulates anterior pituitary hormones, which trigger hormone release by other endocrine glands, as illustrated below.

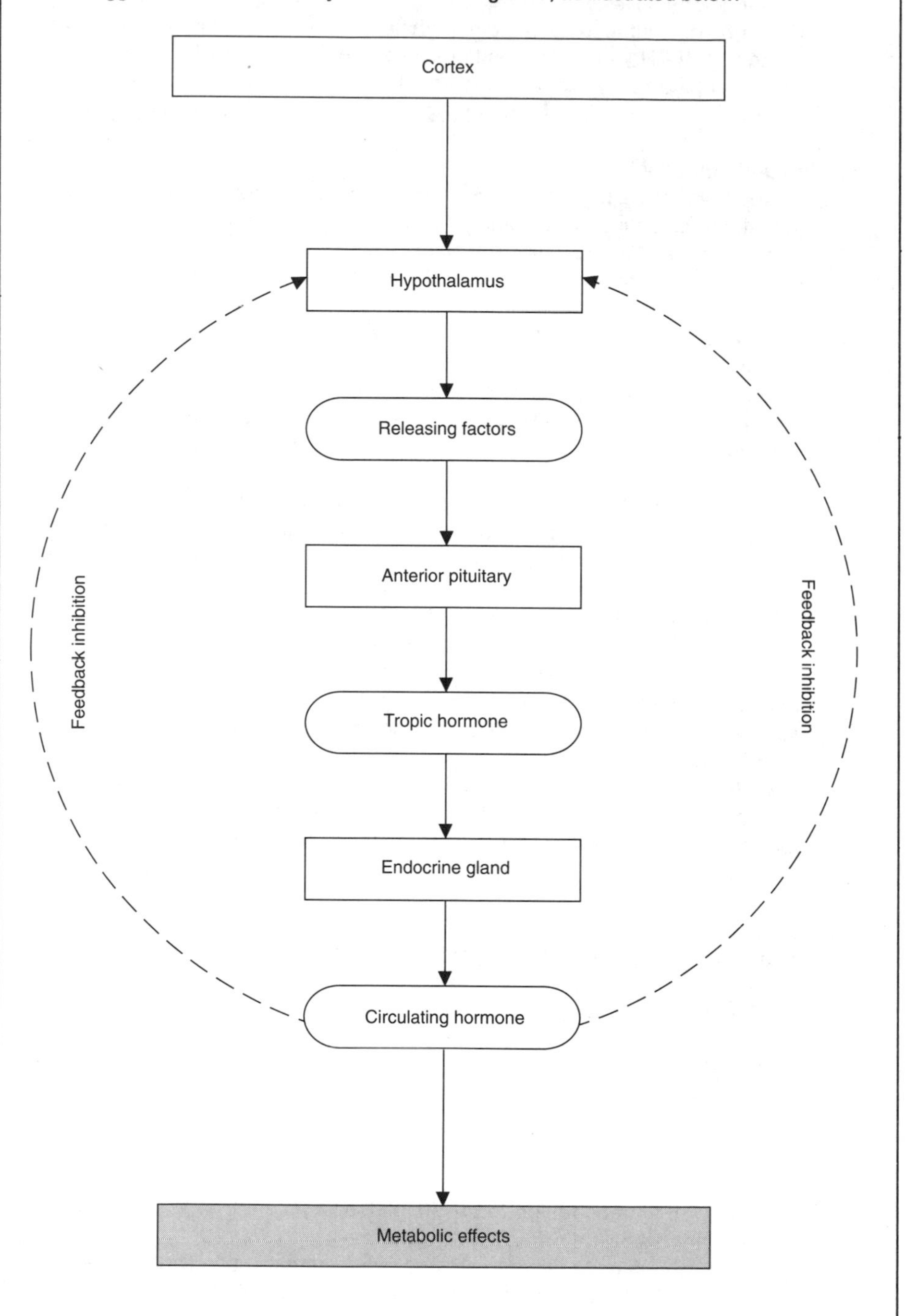

(a) It promotes protein synthesis by facilitating amino acid entry into cells
(b) It helps mobilize fat from adipose tissue
(c) It enhances fat metabolism for energy, which suppresses carbohydrate metabolism; as a result, the blood glucose level rises because glucose metabolism is reduced

(2) Two hypothalamic releasing factors control GH release: growth hormone – releasing hormone and growth hormone – inhibiting hormone, or somatostatin

b. Prolactin stimulates milk secretion by the breasts after they have been prepared by estrogen, progesterone, and other hormones
(1) Unlike other pituitary hormones, prolactin output is suppressed by its hypothalamic inhibitory factor, prolactin inhibiting factor (PIF); prolactin secretion would continue unabated if it were not held in check by PIF
(2) Thyrotropin-releasing hormone (TRH) stimulates prolactin secretion

c. TSH stimulates the thyroid gland to synthesize and secrete thyroid hormones
(1) Thyroid hormones include thyroxine (T_4) and triiodothyronine (T_3); these hormones control general metabolic processes
(2) TRH stimulates TSH release

d. ACTH stimulates the adrenal cortex to manufacture and secrete adrenocorticol hormones
(1) Hormone output is controlled by ACTH-releasing factor, a regulatory hormone produced by the hypothalamus
(2) ACTH exerts its main effect on the adrenal hormones that control carbohydrate metabolism (glucocorticoids)

e. FSH and LH are gonadotropic hormones
(1) These hormones regulate the growth and development of the gonads and control the output of sex hormones that are responsible for the development of male and female secondary sex characteristics (see Chapter 13, Reproductive System, for more information)
(2) Luteinizing hormone – releasing hormone, a releasing factor from the hypothalamus, regulates FSH and LH

D. Posterior lobe

1. The posterior lobe consists of a meshwork of nerve fibers and specialized cells called pituicytes
2. Bundles of nerve fibers in the pituitary stalk connect the posterior lobe to the hypothalamus; these structures are not linked by the portal venous pathways
3. The posterior lobe secretes oxytocin and antidiuretic hormone (ADH)

a. Oxytocin stimulates uterine contractions and causes ***milk ejection*** from the breasts
(1) Oxytocin stimulates contraction of the pregnant uterus
(2) Milk ejection occurs when oxytocin stimulates contraction of specialized contractile cells surrounding the breast glands and ducts
(a) Nipple stimulation from breast-feeding initiates nerve impulses
(b) These impulses are transmitted to neurons in the hypothalamus, which send impulses to the posterior lobe and cause oxytocin release

Pituitary Hormone Functions

Hormones from the anterior and posterior lobes of the pituitary gland perform different functions, as summarized below.

HORMONE	PRINCIPAL FUNCTIONS
Anterior lobe	
Growth hormone (GH, somatotropin)	Accelerates body growth
Prolactin	Stimulates milk secretion
Thyroid-stimulating hormone (TSH, thyrotropin)	Stimulates thyroid hormone synthesis and secretion
Adrenocorticotropic hormone (ACTH, corticotropin)	Stimulates adrenocortical hormone secretion
Follicle-stimulating hormone (FSH)	Stimulates ovarian follicle growth in women and spermatogenesis in men; also regulates gonadal growth and development
Luteinizing hormone (LH)	Stimulates ovulation and luteinization of ovarian follicles in women and testosterone secretion in men; also regulates gonadal growth and development
Posterior lobe	
Oxytocin	Causes milk ejection and stimulates contraction of pregnant uterus
Antidiuretic hormone (ADH, vasopressin)	Promotes water retention

b. ADH causes the cells of the renal and collecting tubules to become more permeable to water, altering the urine concentration (see Chapter 10, Urinary System, for more information)
   (1) Osmoreceptors in the hypothalamus regulate ADH secretion by responding to osmolarity variations in the extracellular fluid (ECF)
   (2) If ECF osmolarity rises above normal, hypothalamic neurons send impulses to the posterior lobe, stimulating ADH release; this causes water retention, diluting the ECF and lowering its osmolarity
   (3) If ECF osmolarity falls below normal, the hypothalamus reduces ADH output from the posterior lobe; this prevents reabsorption of excess water in the tubular filtrate and causes it to be excreted in the urine, concentrating the ECF and increasing its osmolarity (see *Pituitary Hormone Functions* for a summary)

III. Thyroid Gland

A. General information

1. The thyroid gland consists of two lateral lobes connected by a narrow isthmus; it is fixed to the anterior surface of the upper trachea by loose connective tissue
2. This gland is composed of spherical thyroid follicles that contain colloid surrounded by a layer of cuboidal follicular cells
3. Follicular cells synthesize the thyroid hormones T_3 and T_4
   a. These hormones regulate metabolic processes and are required for normal growth and development, especially of the nervous system
   b. The general term *thyroid hormone* refers to T_3 and T_4

4. Parafollicular cells (specialized cells located between the follicles) secrete the hormone calcitonin, which, along with parathyroid hormone, regulates calcium metabolism

B. Thyroid hormone synthesis

1. Follicular cells actively concentrate iodine; they contain enzymes that can synthesize thyroid hormone from iodine and the amino acid tyrosine
2. Iodine circulates in the bloodstream as iodide ions (I^-)
   a. The thyroid gland takes up these ions and oxidizes them to iodine (I_2)
   b. Iodine combines with tyrosine to form monoiodotyrosine (MIT) and diiodotyrosine (DIT)
   c. Condensation of MIT and DIT molecules yields T_3 and T_4
3. T_3 and T_4 combine with the large protein molecule thyroglobulin and form the colloid of the thyroid follicles

C. Thyroid hormone secretion

1. Pituitary TSH controls T_3 and T_4 output, which is regulated by a negative-feedback mechanism
2. Thyroid hormone stored as colloid must be released from thyroglobulin before it can be secreted; thyroid follicular cells perform this function
   a. These cells ingest colloid by pinocytosis
   b. This causes T_3 and T_4 to split off and be secreted into the bloodstream
   c. It also degrades thyroglobulin into its component amino acids, which become part of the amino acid pool
3. Most of the thyroid hormone is metabolically inactive and circulates bound to a plasma protein called thyroid-binding globulin; only the small amount of free (unbound) hormone is metabolically active
4 Most of the secreted hormone is T_4, but T_3 exerts a greater metabolic effect than T_4; much of the T_4 is converted to T_3 in the tissues where it exerts its effect

D. Metabolic effects of thyroid hormone

1. Thyroid hormone controls the rate of metabolic processes
2. It is required for normal growth and development and for development and maturation of the nervous system
3. A thyroid hormone excess (hyperthyroidism) or deficiency (hypothyroidism) leads to a corresponding increase or decrease in metabolic processes

IV. Parathyroid Glands

A. General information

1. Four small parathyroid glands are embedded in the surface of the thyroid gland
2. These glands secrete parathyroid hormone, the principal regulator of calcium metabolism
3. Blood calcium levels normally range from 8.5 to 10.5 mg per 100 ml of blood
4. Half the calcium in blood is in ionized form as Ca^{++}, which is the active form; the other half is bound to blood proteins and is biologically inactive
5. An adequate level of ionized calcium is required for normal cardiac and skeletal muscle contraction, nerve impulse transmission, and blood coagulation; too lit-

tle ionized calcium greatly increases nerve and muscle excitability, whereas too much ionized calcium diminishes it

6. A reciprocal relationship exists between the levels of calcium and phosphorus in the blood
   a. A decreased calcium level tends to increase the phosphorus level
   b. An increased calcium level tends to decrease the phosphorus level
7. The average diet contains 1 to 2 grams of calcium daily, but only about 10% is absorbed
   a. Milk and dairy products are the chief sources of dietary calcium
   b. Phosphate is abundant in the average diet and is absorbed readily with calcium
8. Parathyroid hormone, vitamin D, and the thyroid hormone calcitonin regulate the blood calcium level
   a. Parathyroid hormone and vitamin D raise the blood calcium level
   b. Calcitonin lowers the blood calcium level
9. Because calcium in the blood is in equilibrium with calcium salts in bone, changes in blood calcium level eventually cause changes in the amount of calcium salts in the bone

B. Calcium metabolism regulators

1. Parathyroid hormone, a protein (polypeptide) hormone, helps regulate calcium metabolism
   a. This hormone is not stored, but is synthesized and secreted continuously
   b. The ionized calcium level in the blood regulates parathyroid hormone output by a negative feedback mechanism
      (1) A decrease in the ionized calcium level causes increased parathyroid hormone output, which raises the blood calcium level
      (2) An increase in the ionized calcium level suppresses parathyroid hormone secretion, which reduces the blood calcium level
   c. Parathyroid hormone has three sites of action: the skeletal system, intestines, and kidneys
      (1) Its main function is to mobilize calcium from bone by promoting bone matrix breakdown and liberating calcium, which diffuses into the blood and raises the blood calcium level
      (2) In the intestines, parathyroid hormone increases calcium absorption, which tends to raise the blood calcium level
      (3) In the kidneys, the hormone increases calcium reabsorption by the renal tubules and promotes phosphate excretion
         (a) Increased phosphate excretion in urine lowers the phosphate concentration in blood
         (b) Because calcium and phosphate blood levels have a reciprocal relationship, the calcium level rises correspondingly
2. Vitamin D also regulates calcium metabolism
   a. This vitamin promotes calcium absorption from the intestines
   b. It is formed by a complex process from a cholesterol derivative in the skin
      (1) Ultraviolet light in sunlight converts this sterol into an intermediate compound
      (2) The liver and kidneys further metabolize the intermediate compound into the active form of vitamin D
3. Calcitonin acts as an antagonist to parathyroid hormone

a. Calcitonin is a polypeptide hormone formed by the parafollicular cells of the thyroid gland
b. This hormone is secreted in response to an increased blood calcium level; it tends to lower the blood calcium level, primarily by inhibiting calcium mobilization from bone

V. Adrenal Glands

A. General information

1. The adrenal glands are located on top of the kidneys; each consists of an outer cortex, which is enclosed in a fibrous capsule, and an inner medulla
2. These two distinct parts of the glands secrete different hormones
   a. The adrenal cortex secretes three types of steroid hormones: glucocorticoids, mineralocorticoids, and sex hormones
   b. The adrenal medulla produces two similar hormones: norepinephrine and epinephrine

B. Adrenal cortex glucocorticoids

1. The major glucocorticoid is cortisol (hydrocortisone); others include corticosterone and cortisone
2. Cortisol and other glucocorticoids have similar actions
   a. They raise the blood glucose level by decreasing glucose metabolism and by promoting glucose formation from protein and fat (gluconeogenesis)
   b. They promote protein breakdown into amino acids, some of which are converted by the liver into glucose; this depletes tissue proteins, which are converted to glucose
3. Glucocorticoids are secreted in response to ACTH stimulation
4. Their output is controlled by a negative feedback mechanism in which a low glucocorticoid level stimulates ACTH secretion and a high level suppresses it

C. Adrenal cortex mineralocorticoids

1. Mineralocorticoids regulate electrolyte and water balance by promoting sodium ion (Na^+) absorption and potassium ion (K^+) excretion by the renal tubule (see Chapter 10, Urinary System, for more information)
2. The major mineralocorticoid is aldosterone; its secretion is regulated by more than one mechanism
   a. Although ACTH increases aldosterone secretion somewhat, the most potent stimulus is the renin-angiotensin-aldosterone system
   b. This system responds to variations in blood volume, blood pressure, and blood sodium concentration; its effect is mediated by the juxtaglomerular apparatus of the kidneys

D. Adrenal cortex sex hormones

1. Small amounts of estrogen, progesterone, and testosterone are produced by the adrenal glands of both sexes
2. The amounts produced are minimal compared with the much larger quantities produced by the gonads
3. The small amount of androgen produced by the adrenal glands appears to be responsible for sex drive in women

E. Adrenal medulla hormones

1. The adrenal medulla produces the hormones norepinephrine and epinephrine from a precursor amino acid called tyrosine
2. Both hormones belong to a class of compounds called catecholamines; epinephrine differs from norepinephrine only in the presence of a methyl group (CH_3^-) attached to the amino group in the molecule
3. During synthesis, norepinephrine is formed first; then some of it is converted to epinephrine by the addition of the methyl group
4. Both hormones are stored in the cytoplasmic granules of adrenal medulla cells; they are released into circulation in response to nerve impulses transmitted by preganglionic fibers of the sympathetic nervous system
5. Of the hormones produced in the adrenal medulla, norepinephrine accounts for about 20%; epinephrine accounts for the remainder
6. Emotional stress, such as anger, fear, or anxiety, activates the sympathetic nervous system and causes the adrenal medulla to release these hormones
7. The liberated catecholamines exert widespread effects that produce a physiologic response to stress —the fight-or-flight response
   a. This response increases the heart rate and cardiac output and constricts the blood vessels
   b. It elevates the blood glucose level and mobilizes glycogen from the liver and free fatty acids from adipose tissue to provide glucose for energy
   c. This response also increases the responsiveness of the nervous system
8. Small amounts of catecholamines are excreted unchanged in the urine, but most are inactivated by various enzymes and excreted in an inactive form
9. Inactivation is accomplished in one of two ways
   a. A methyl group can be added to one of the hydroxyl groups in the catecholamine ring, forming a compound called a metanephrine
   b. The amino group can be removed, the terminal carbon oxidized to a carboxyl group, and a methyl group added to one of the hydroxyl groups attached to the ring; this produces vanillylmandelic acid (VMA)

VI. Islets of Langerhans

A. General information

1. The islets are about one million cell clusters scattered throughout the pancreas
2. Each islet is composed of three major cell types and three minor cell types
   a. The major types are classified as alpha, beta, or gamma cells
   b. The minor types are classified as pancreatic polypeptide (PP), D_1, or enterochromaffin cells
3. Alpha cells, which constitute 20% of the islet cells, secrete glucagon; this raises the blood glucose level by promoting the conversion of liver glycogen into glucose, which is liberated into the circulation
4. Beta cells constitute 70% of islet cells; they secrete insulin (the major islet hormone), which lowers the blood glucose level

5. Delta cells make up 5% to almost 10% of islet cells; they produce a hormone called somatostatin, which suppresses insulin and glucagon release from the islets
6. PP cells make up 1% to 2% of islet cells; they produce a hormone that stimulates GI enzyme secretion and inhibits intestinal motility
7. Only small numbers of D_1 and enterochromaffin cells can be identified in normal islets
   a. D_1 cells produce a hormone called vasoactive intestinal polypeptide, which increases the blood glucose level and stimulates GI secretions
   b. Enterochromaffin cells synthesize serotonin
8. Insulin is the islet cell hormone of major physiologic importance

B. Insulin

1. Within the beta cell, insulin is synthesized by the endoplasmic reticulum and then transported to the Golgi apparatus
2. Next, insulin is formed into secretory granules that accumulate in the cytoplasm of the beta cells and eventually are discharged into the circulation
3. Insulin is synthesized as a large precursor peptide molecule called proinsulin, which is largely inactive
   a. Chemical bonds join the two ends of the peptide chain
   b. Each bond consists of a cross-bridge made of two sulfur atoms (disulfate bond); cross-linkage causes the chain to coil
4. As the proinsulin is being stored as secretory granules, a trypsin-like enzyme from Golgi apparatus membranes splits off the central part of the proinsulin coil to yield the active enzyme insulin and the connecting peptide
   a. Insulin consists of two short peptide chains joined together by disulfide bonds
   b. Both cleavage parts are stored in the secretory granules and secreted together into the bloodstream

C. Insulin secretion

1. Insulin secretion occurs in two phases
   a. An initial outpouring results from the discharge of insulin stored in the cytoplasmic granules of the beta cells
   b. A slower, more sustained release follows; this results from synthesis and release of additional insulin by beta cells
2. The main stimulus for insulin secretion is blood glucose elevation, which occurs after eating
3. Ingested glucose causes much more insulin to be secreted than does the same amount of intravenous glucose, because digestion triggers the release of hormones and amino acids that also stimulate insulin release
4. Several other hormones increase the blood glucose level, indirectly stimulating insulin secretion
   a. Glucagon and catecholamines raise the blood glucose concentration by promoting conversion of liver glycogen into glucose
   b. GH inhibits glucose use by the tissues; it promotes fat breakdown to yleld free fatty acids, which are used for energy instead of glucose
   c. Glucocorticoids raise the blood glucose level primarily by promoting protein breakdown into amino acids, which are converted into glucose by the liver (gluconeogenesis)

D. Actions of insulin

1. Insulin has multiple effects, influencing not only carbohydrate metabolism but protein and fat metabolism as well
2. The chief sites of insulin action are liver cells, muscle tissue, and adipose tissue
    a. Insulin promotes the entry of glucose into the cells and enhances the use of glucose as a source of energy
    b. It promotes the storage of glucose as glycogen in muscle and liver cells; in adipose tissue, insulin enhances the conversion of glucose to triglyceride and storage of the newly formed triglyceride within fat cells
    c. Insulin also promotes the entry of amino acids into cells and stimulates protein synthesis

---

Study Activities

1. Compare the functions of exocrine and endocrine glands.
2. Describe how hormones exert intracellular effects by binding with receptors on the exterior of the cell.
3. Trace the negative feedback mechanism by which the hypothalamus controls pituitary gland secretion.
4. Identify the major anterior and posterior pituitary hormones and discuss their functions.
5. Discuss the steps involved in thyroid hormone storage and synthesis.
6. Compare the roles of parathyroid hormone, vitamin D, and calcitonin in calcium metabolism regulations.
7. Describe the functions of the adrenal cortex glucocorticoids, mineralocorticoids, and sex hormones and the adrenal medulla hormones.
8. Discuss the secretion and actions of insulin.

13

Reproductive System

Objectives

After studying this chapter, the reader should be able to:

- Identify the structures of the male reproductive system and explain their functions.
- Explain how gonadotropic and testicular hormones regulate spermatogenesis and testosterone production.
- Discuss the structure and function of spermatozoa.
- Describe the composition and functions of semen.
- Trace the sequence of events in the male sexual response.
- Identify the structures of the female reproductive system and explain their functions.
- Explain the steps of oogenesis from primary oocyte to ovum.
- Describe the normal menstrual cycle, identifying the effects of the gonadotropic and ovarian hormones that regulate it.
- Trace the sequence of events in the female sexual response.
- Describe the events that occur during puberty and cause mature gamete production.
- Compare the decline of gonadal function in men and women.

I. Male Reproductive System

A. General information

1. The male reproductive system consists of the penis, scrotum, testes, a duct system, and accessory organs
    a. The penis is the copulatory organ, by which spermatozoa are placed in the female reproductive system
        (1) It consists of three cylinders of vascular erectile tissue called cavernous bodies
            (a) The upper (dorsolateral) two cylinders, the corpora cavernosa, are encased by dense fibrous tissue
            (b) The lower (midventral) one, the corpus spongiosum, encloses the urethra
        (2) Three pairs of superficial skeletal muscles attach to the bases of the cavernous bodies; contraction of these muscles compresses the urethra, which occurs during ***ejaculation*** and urination
    b. Suspended below the penis, the scrotum covers and protects the testes and spermatic cords and maintains the testes at the proper temperature for spermatozoa production
    c. The testes, which are located in the scrotum, are the male gonads

  (1) They produce spermatozoa in the seminiferous tubules
  (2) They also produce the hormone testosterone in *Leydig's cells* (interstitial cells between the seminiferous tubules)
  (3) Gonadotropic hormones regulate testicular functions
d. The duct system moves spermatozoa from the testes to outside the body
  (1) The rete testis and efferent ducts lie within each testis and drain into the epididymis, which is located on top of the testis
  (2) The duct system continues as the vas deferens, which extends to the prostate gland and joins the seminal vesicles to form the ejaculatory duct; this duct empties into the urethra
e. The accessory glands, which include the seminal vesicles, prostate gland, and bulbourethral (Cowper's) glands, and secrete the liquid portion of semen
  (1) The seminal vesicles merge with the vas deferens and secrete a thick alkaline fluid that nourishes spermatozoa and enhances their motility
  (2) The prostate gland surrounds the urethra and lies below the bladder; it secretes a thin alkaline fluid, which promotes sperm motility and is discharged into the urethra through small ducts that open near the orifices of the ejaculatory ducts
  (3) Cowper's glands, located below the prostate, secrete a mucoid substance that provides lubrication during intercourse

2. The male reproductive system produces semen, a mixture of spermatozoa and various fluids; semen discharge is necessary for fertilization and involves a two-phase spinal reflex

B. Hormonal regulation of testicular function

1. Two gonadotropic hormones regulate testicular function: *follicle-stimulating hormone (FSH)* and *luteinizing hormone (LH)*
   a. The anterior pituitary gland releases FSH and LH
   b. The hypothalamus controls FSH and LH secretion, which is continuous in males
   c. A negative feedback mechanism controls FSH and LH output; rising FSH and LH levels suppress further output of these hormones, which maintains stable gonadotropic hormone levels
2. FSH has two functions
   a. It promotes development and normal function of the seminiferous tubules
   b. FSH helps stimulate spermatozoa production
3. Inhibin suppresses FSH release; this protein (polypeptide) hormone is produced by specialized cells of the seminiferous tubules called Sertoli's cells
   a. Inhibin levels rise with active ***spermatogenesis;*** elevated inhibin levels suppress further FSH output
   b. This prevents excessive spermatogenesis and maintains normal sperm production
4. LH also promotes testosterone secretion from Leydig's cells
   a. Testosterone is responsible for sexual drive, development of secondary sex characteristics, and growth
   b. It also promotes normal spermatogenesis in the seminiferous tubules
5. Rising testosterone levels suppress further LH output, thereby maintaining a stable testosterone level
6. Spermatogenesis and testosterone production are separate functions

C. Spermatogenesis

1. Spermatogenesis is the process of spermatozoa formation
2. Precursor cells in the seminiferous tubules called *spermatogonia* contain 46 chromosomes
3. Spermatogonia divide repeatedly by mitosis to form *primary spermatocytes,* which also contain 46 chromosomes
4. Primary spermatocytes then undergo meiotic divisions; the number of chromosomes in the cells is reduced by half (see Chapter 1, Human Cell, for more information on meiosis)
   a. In the first meiotic division, each primary spermatocyte forms two *secondary spermatocytes* (each with 23 chromosomes)
   b. In the second meiotic division, each secondary spermatocyte forms two *spermatids* (each with 23 chromosomes), which mature into spermatozoa (with 23 chromosomes)
5. Spermatozoa are produced continuously in the testes and seminiferous tubules; the production process takes about 2 months
6. LH and FSH are needed to maintain normal spermatogenesis
   a. LH stimulates Leydig's cells to secrete testosterone, which is required for spermatogenesis
   b. FSH stimulates Sertoli's cells in the seminiferous tubules to secrete androgen-binding protein
      (1) This protein binds with the secreted testosterone, maintaining a high hormone level at the site of spermatogenesis
      (2) Sertoli's cells also provide nutrients to maturing spermatids, which attach themselves to these support cells

D. Spermatozoa structure and function

1. The spermatozoon is a tadpole-like structure that consists of three parts: head, middle piece, and tail
   a. The head, which contains the chromosomes, is partially covered by a thin membrane-like structure called the head cap or acrosome; the acrosome contains enzymes that allow the spermatozoon to penetrate and fertilize the ovum
   b. The middle piece contains mitochondria with enzymes that provide the energy required to propel the spermatozoon
   c. The tail propels the sperm by flagellated movement (to and fro motion)
2. Spermatozoa discharged in semen must undergo activation via ***capacitation;*** this involves a structural change in the spermatozoon
   a. Small perforations appear in the acrosome of the spermatozoon
   b. These perforations allow the release of enzymes required for the spermatozoon to penetrate the ovum

E. Semen composition

1. Semen is a viscous secretion that consists of spermatozoa and secretions from the seminal vesicles and the prostate and Cowper's glands
   a. Seminal vesicles and prostate secretions contribute most of the semen volume and perform two functions
      (1) Their slight alkalinity protects spermatozoa from the acidity of the vaginal secretions
      (2) They provide nutrients for spermatozoa

   b. Cowper's glands provide some lubricating fluid
2. The average volume of ejaculated semen is about 3 ml; the volume normally varies from 2 to 5 ml, depending on the time interval between ejaculations
3. Normally, 1 ml of semen contains up to 100,000,000 spermatozoa
   a. The spermatozoa are not distributed uniformly in semen
   b. Most are present in the first part of the ejaculate
4. Although only one spermatozoon can enter the ovum to fertilize it, large numbers of spermatozoa are ejaculated to help ensure that one survives to achieve ***fertilization;*** a spermatozoa count under 20,000,000 per ml of semen usually is associated with sterility

F. Male sexual response

1. In men, ***sexual response*** consists of penile erection and semen discharge
2. The penile erectile tissue consists of three cylinders composed of a spongy meshwork of endothelium-lined blood sinuses supplied by many arterioles and drained by veins
3. The erectile tissue cylinders, surrounded by fibrous capsules, are called cavernous bodies
4. Because the arterioles that supply the erectile tissue of the cavernous bodies normally are contracted, little blood flows into the cavernous bodies, and the sinusoids are collapsed
5. Sexual excitement causes reflex dilation of the arterioles; this results from parasympathetic nerve stimulation and concomitant inhibition of sympathetic nerves, which normally constrict arterioles
   a. The cavernous bodies become engorged with blood, and the veins that drain them become compressed, causing the penis to become rigid
   b. The penis returns to the flaccid state as a result of parasympathetic nerve inhibition and sympathetic nerve stimulation; this causes the arterioles to constrict again and allows the excess blood to drain from the cavernous bodies
6. Semen discharge involves a two-phase spinal reflex, consisting of ***emission*** and *ejaculation*
7. Emission refers to semen movement into the urethra
   a. Emission occurs when the sympathetic nervous system transmits efferent impulses from the lumbar spinal cord
   b. These impulses cause rhythmic contractions of the muscles near the epididymis, vas deferens, seminal vesicles, and prostate gland to expel semen into the urethra
8. Ejaculation is the pulsatile expulsion of semen from the penis; it is associated with erotic sensations called ***orgasm***
   a. Ejaculation occurs when efferent impulses from the sacral portion of the spinal cord cause rhythmic contractions of the muscles around the base of the penis
   b. This intermittently compresses the urethra, causing semen to be expelled from the penis in spurts
9. During emission and ejaculation, the sphincter muscle at the base of the bladder constricts, preventing semen reflux into the bladder
10. The two components of the ejaculatory spinal reflex are integrated but independent functions; for example, the emission phase can be abolished by surgical removal of the sympathetic chains that convey sympathetic impulses to the

muscles near the epididymis and ducts that transport sperm, but the muscle contractions —and orgasm —of the ejaculation phase will remain

11. The two-phase spinal reflex occurs when the impulses from the penis and genital region, other skin areas, and cerebral cortex reach a critical level of intensity

II. Female Reproductive System

A. General information

1. The female reproductive system includes the external and internal genitalia
2. The external genitalia, or *vulva,* include the mons pubis, labia majora, labia minora, clitoris, and vestibule
   a. The mons pubis is a cushion of adipose and loose connective tissue over the symphysis pubis
   b. The labia majora are twin folds of adipose and connective tissue that run from the mons pubis to the perineum
   c. The labia minora are twin folds of connective tissue between the labia majora
   d. Similar to the penis, the clitoris is composed of erectile tissue; it is surrounded by mucosal folds (the prepuce of the clitoris)
   e. The vestibule is an oval-shaped structure surrounded by the clitoris and labia; several internal structures open into the vestibule, including the urethra, vagina, and ducts from Skene's and Bartholin's glands, which secrete lubricating substances during intercourse
3. The internal genitalia include the vagina, cervix, uterus, fallopian tubes, and ovaries
   a. The vagina is a fibromuscular tube that extends from the vestibule to the cervix
      (1) It serves as a passageway for menstrual flow and as the receptacle for the penis during intercourse
      (2) It also is the lower part of the birth canal
   b. The cervix is the narrow inferior portion of the uterus
      (1) It projects into the upper end of the vagina
      (2) The cervix is surrounded by the vaginal fornices (recesses in the vaginal wall)
   c. The uterus is a pear-shaped structure with a thick muscular wall (myometrium); it is designed to receive a fertilized ovum and support fetal development
      (1) The organ is lined by a glandular mucous membrane (endometrium)
      (2) It is held in position by bands of connective tissue called ligaments
   d. The fallopian tubes extend laterally from the upper corners of the uterus
      (1) These tubes transport ova from the ovaries to the uterus
      (2) The fringed (fimbriated) ends partially surround the adjacent ovaries
   e. The ovaries are the female gonads; they produce ova
4. ***Oogenesis*** recurs periodically; in this process, each diploid precursor cell produces one hapoid mature ova and three haploid polar bodies
5. The normal female reproductive cycle, or ***menstrual cycle,*** involves interaction of pituitary gonadotropic hormones and ovarian hormones, which prepare the endometrium to receive the fertilized ovum; if fertilization and implantation do not occur, the prepared endometrium is discarded through menstruation and a new cycle starts

6. Variations in menstrual cycle duration primarily result from variations in preovulatory phase duration; the postovulatory phase remains relatively constant at about 14 days
7. Female sexual response resembles the male response, except that it does not produce a discharge comparable to ejaculate

B. Menstrual cycle hormones

1. Pituitary and ovarian hormones induce cyclic changes in the endometrium that are responsible for the menstrual cycle
2. These hormones also produce less pronounced cyclic changes in the breasts, fallopian tubes, and cervical mucosa
3. The pituitary gonadotropic hormones FSH and LH are secreted cyclically in women
   a. FSH stimulates the ovarian follicles to grow and secrete estrogen
   b. LH acts with FSH to promote follicle maturation
      (1) Under the influence of FSH and LH, a follicle develops and discharges its ovum; this action marks ***ovulation***
      (2) Then LH causes the ruptured follicle to change into a large convoluted yellow structure called the corpus luteum, which produces the steroid hormones estrogen and progesterone
4. Under the influence of FSH and LH, the ovary secretes estrogen and progesterone
5. The estrogen secreted by the ovaries exerts several effects
   a. The hormone induces sexual development at ***puberty,*** including proliferation of glandular tissue of the breasts
   b. It stimulates endometrial growth during the first half (proliferative phase) of the menstrual cycle
   c. It stimulates the glandular epithelium of the cervix to secrete a thin, alkaline mucus that facilitates spermatozoa passage into the uterus and fallopian tube
   d. Estrogen also has general metabolic effects, such as stimulation of bone growth
6. Progesterone, secreted by the corpus luteum (a yellow globular mass in the ovaries), produces several effects
   a. It induces marked secretory activity in the endometrial glands during the second half (secretory phase) of the menstrual cycle, preparing the endometrium for implantation of the fertilized ovum
   b. The hormone also increases cervical mucus thickness and viscosity, making it relatively resistant to spermatozoa penetration during the postovulatory phase
   c. Progesterone also causes breast development
7. Gonadotropic and ovarian hormones have a reciprocal relationship
   a. A high estrogen level inhibits FSH output, stabilizing the estrogen level; it also stimulates LH release, which increases progesterone output
   b. A high progesterone level inhibits LH output

C. Oogenesis

1. Oogenesis is the process of ova formation
2. Precursors of the ova, or *oogonia,* proliferate by mitosis in the fetal ovaries before birth; they form *primary oocytes* (with 46 chromosomes)

3. Then a single layer of granulosa or follicular cells surrounds the oocytes, forming structures called *primary follicles*
   a. The primary oocytes in the primary follicles enter the prophase of the first meiotic division during fetal development, but they do not continue to divide
   b. A very large number of primary follicles form
      (1) Many degenerate from infancy to childhood, but about 500,000 remain into adolescence
      (2) This number is far more than required; less than 500 ova are released in a lifetime, and only a few of these are fertilized
4. The primary follicles remain inactive until puberty, when cyclic ovulation begins under the influence of FSH and LH
5. During each menstrual cycle, primary follicles begin to grow in the ovary, but normally only one matures and is ovulated
   a. The granulosa cells around the follicles proliferate, and a layer of noncellular material called the zona pellucida is deposited on the surface of the oocyte
   b. Fluid begins to accumulate in the layer of granulosa cells; a central fluid-filled cavity forms in the mature or graafian follicle
6. About the time the oocyte is released at ovulation, it completes its first meiotic division, giving rise to two daughter cells —a secondary oocyte and the first polar body, which are of unequal size
   a. The secondary oocyte contains half the chromosomes (23) and almost all the cytoplasm
   b. The first polar body contains the remaining 23 chromosomes but almost no cytoplasm
7. The newly formed secondary oocyte begins its second meiotic division, but it does not complete this division unless it is fertilized; completion of the second division gives rise to a mature ovum and a second polar body, each containing 23 chromosomes
   a. Whether fertilization occurs or not, the first polar body immediately undergoes a second meiotic division, giving rise to two additional haploid polar bodies (for a total of three), which degenerate
   b. The ovum can survive for about 3 days after ovulation; however, it can only be fertilized successfully for about 36 hours after ovulation
8. The ovum is swept into the fimbria of the fallopian tube by the beating of cilia that cover the tubal epithelium; it is propelled down the fallopian tube by the cilia and by peristaltic contractions of smooth muscle in the wall of the tube
9. Ova released late in a woman's reproductive life have been arrested in prophase for up to 45 years before resuming meiosis at ovulation; this may explain the high incidence of congenital abnormalities related to abnormal chromosome separation in the offspring of older women

D. Menstrual cycle

1. The beginning of the cycle customarily is dated from the first day of menstrual flow
2. Ovulation occurs around the middle of the cycle, dividing it into preovulatory (follicular) and postovulatory (luteal) phases
3. Menstrual cycle duration varies considerably among individuals; it also may vary somewhat from month to month in an individual (see *Events in the Menstrual Cycle,* page 155, for diagrams)
   a. Variations usually result from differences in the preovulatory phase duration

   b. The duration of the postovulatory phase remains relatively constant at about 14 days
4. During the preovulatory phase, the pituitary gland begins to release FSH, which stimulates a group of ovarian follicles to grow
   a. The first follicle to respond grows more rapidly and comes to full maturity
   b. The other follicles undergo involution (atrophy)
5. Soon after FSH output rises, LH output begins to increase; together FSH and LH promote estrogen secretion by the ovarian follicles
6. The increasing estrogen output inhibits further FSH release and stimulates LH release
   a. LH release eventually builds to a precipitous outpouring called the LH surge, which persists for about 24 hours
   b. This surge leads to rupture of the follicle and ovulation
7. The postovulatory phase is characterized by conversion of the ruptured follicle into a corpus luteum, which produces estrogen and progesterone
8. The corpus luteum reaches maturity about 8 to 9 days after ovulation; then it begins to degenerate if pregnancy has not occurred
9. Menstruation results from the decline in corpus luteum activity, which reduces estrogen and progesterone output
   a. Eventually, the levels of these two hormones no longer can maintain the endometrium
   b. Then the endometrium is shed, producing the menstrual flow
   c. Menstruation lasts about 5 days; it is associated with an average total blood loss of 50 to 150 ml
10. As estrogen and progesterone levels fall, their inhibitory effects on FSH and LH release also decline
11. Eventually, when estrogen and progesterone have fallen to low levels, FSH and LH are released again and a new cycle begins

E. Female sexual response

1. The physiologic effects of sexual excitement in women are similar to those in men
2. The parasympathetic nervous system controls the dilation of arterioles that supply the erectile tissue in the clitoris and labia minora
   a. During sexual excitement, nerve impulse transmission causes the tissues to become engorged with blood
   b. Parasympathetic impulses also cause Skene's and Bartholin's glands to produce secretions that lubricate the vagina during intercourse
3. Sensory input from nerve endings in the clitoris, labia, vaginal orifice, and perineal tissues is transmitted to the lumbar and sacral spinal cord
4. When this input reaches a critical level, a spinal reflex response occurs, characterized by rhythmic contractions of the uterus and fallopian tubes; these contractions, together with rhythmic and pulsatile contractions of the skeletal muscles around the vagina, comprise the erotic sensations called orgasm
5. Female sexual climax differs from male climax in several ways
   a. No secretions are discharged comparable to the ejaculate in males
   b. No sexual climax is required to achieve fertilization

Events in the Menstrual Cycle

The menstrual cycle may be divided into an ovarian cycle with two phases (follicular, or preovulatory, and luteal, or postovulatory) or an endometrial cycle with three phases (menstrual, proliferative, and secretory). The illustration below depicts the relationship of pituitary hormones (luteinizing hormone [LH] and follicle-stimulating hormone [FSH]) and ovarian hormones (estrogen and progesterone) to various phases of the ovarian and endometrial cycles.

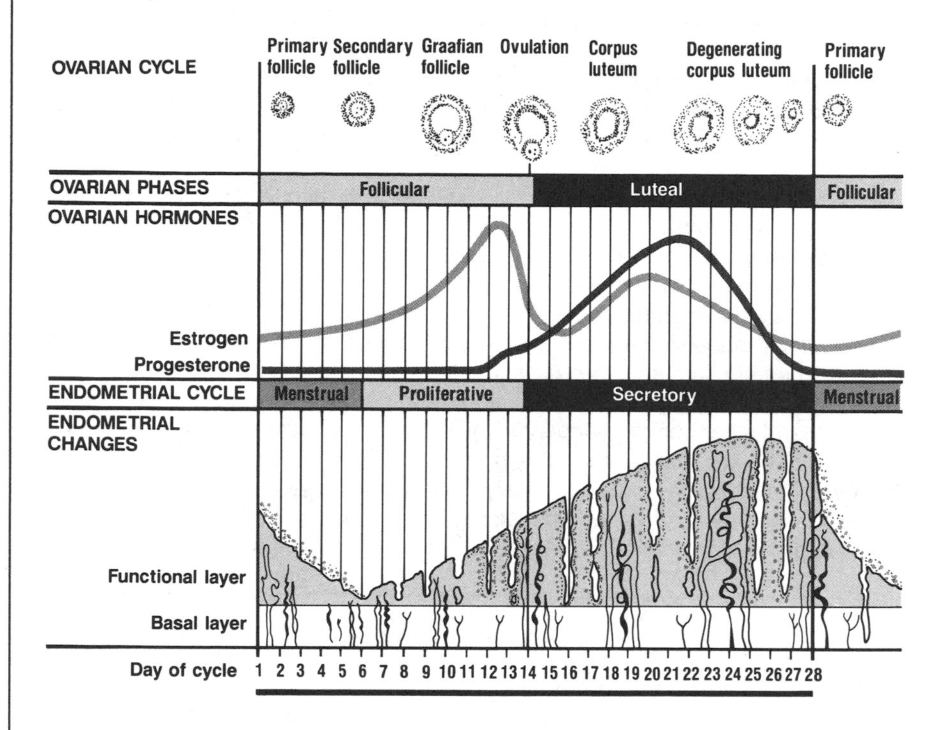

Adapted from Cohen, S., Kenner, C., and Hollingsworth, A. (1991). *Maternal, Neonatal, and Women's Health Nursing.* Springhouse, PA: Springhouse Corp.

III. Onset and Decline of Gonadal Function

A. General information

1. In men and women, the production of mature gametes (spermatozoa or ova) marks *puberty,* the developmental stage in which reproductive ability begins and secondary sex characteristics develop
2. Gonadal function eventually stops in women; in men, gonadal function declines but does not stop

B. Onset of gonadal function

1. Before puberty, trace amounts of sex hormones produced by the gonads are sufficient to suppress production of gonadotropin-releasing hormones by the hypothalamus

2. At puberty, the hypothalamus matures and, for unknown reasons, loses its extreme sensitivity to the inhibitory effect of low sex hormone levels
3. Then the hypothalamus begins to release gonadotropin-releasing hormones; these cause the release of pituitary FSH and LH, which stimulate the gonads to release sex hormones (testosterone in men, estrogen and progesterone in women)
4. Sex hormones induce sexual development and other body changes characteristic of sexual maturity, including ***menarche*** in women

C. Decline of gonadal functions

1. With age, women's ovarian follicles degenerate without producing mature egg cells; mature egg cell production declines until about age 45, when few follicles remain
    a. The ovaries no longer respond to FSH and LH stimulation and stop producing sex hormones
    b. When the ovaries no longer can produce sufficient hormones to stimulate cyclic changes in the endometrium, menstruation ceases; this condition is called ***menopause***
    c. Insufficient sex hormones are produced to inhibit FSH and LH secretion; gonadotropin output then rises
    d. Lack of estrogen causes involution of the female reproductive tract, breasts, and other hormone-dependent tissues; it also can cause vasomotor instability (hot flashes) and other physical and emotional changes
2. In men, testosterone secretion slowly declines with age
    a. Sex drive may decline with testosterone secretion
    b. Spermatogenesis continues, although it may decline somewhat with age

---

Study Activities

1. Contrast spermatogenesis with oogenesis.
2. Trace the flow of spermatozoa through the male reproductive system.
3. Identify the components of normal semen.
4. Compare male and female sexual responses.
5. Trace the movement of an ovum through the female reproductive system.
6. Explain the events of the ovarian and endometrial phases of the menstrual cycle.
7. Characterize the physiologic events that lead to puberty.
8. Explain the causes and effects of menopause.

14

Fertilization, Pregnancy, Labor, and Lactation

Objectives

After studying this chapter, the reader should be able to:

- Identify the basic prerequisites for fertilization, and describe the fertilization process.
- Discuss the changes that occur during the three stages of prenatal development.
- Explain the functions of the corpus luteum during pregnancy.
- Identify the parts of the decidua and describe its functions.
- Explain how amniotic fluid is formed and how maternal and fetal factors regulate its volume.
- Describe the functions of the placenta and the hormones it produces.
- Discuss the factors that contribute to the onset and maintenance of labor.
- Describe the events that occur in each stage of labor.
- Explain the physiologic changes associated with the postpartal period.
- Discuss the initiation and effects of lactation.

I. Fertilization

A. General information

1. *Fertilization* is the union of a *spermatozoon* and an *ovum*
2. Fertilization occurs only when several basic prerequisites are met
    a. Spermatozoa must be adequate in function and number
    b. A mature ovum must be available and ready to be fertilized
    c. Spermatozoa must be transported effectively through the female reproductive tract

B. Spermatozoa transport

1. Spermatozoa move through the female reproductive tract by two mechanisms
    a. They travel several millimeters per hour by using flagellar movement to propel themselves
    b. They are transported up into the uterus and fallopian tubes by rhythmic contractions of uterine muscles
2. Although several hundred million spermatozoa are deposited by a single ejaculation, many are destroyed by the acidity of vaginal secretions
3. Only the spermatozoa that enter the cervical canal, where they are protected by cervical mucus, can survive
4. The ease with which spermatozoa can penetrate cervical mucus is related to the menstrual cycle phase

a. Early in the cycle, spermatozoa have difficulty passing through the cervix because estrogen and progesterone levels cause the mucus to thicken
b. During midcycle, spermatozoa can pass readily through the cervix because the mucus is relatively thin
b. Later in the cycle, spermatozoa have difficultly passing through the cervix because the mucus is much thicker

5. Once through the mucus, the spermatozoa enter the uterus
   a. Only spermatozoa (not seminal fluid) enter the uterus and fallopian tubes
   b. Uterine contractions help spermatozoa ascend into the fallopian tubes
6. Spermatozoa probably can fertilize the ovum for up to 2 days after ejaculation, although they may survive for 3 to 4 days in the reproductive tract

C. Ovum fertilization

1. Fertilization normally occurs in the distal part of the fallopian tube; it is possible when intercourse occurs reasonably close to the time of ovulation based on the viability of the germ cells
   a. The ovum remains viable for 24 to 36 hours
   b. The spermatozoa remain viable for 48 hours or more
2. Before a spermatozoon can penetrate the ovum, it must disperse the granulosa cells and penetrate the zona pellucida; enzymes in the acrosome (head cap) of the spermatozoon allow this penetration
3. Once the spermatozoon has penetrated, the ovum completes its second meiotic division; the zona pellucida becomes impermeable to penetration by other spermatozoa
4. The spermatozoon head fuses with the ovum nucleus, forming a cell nucleus with 46 chromosomes
5. The fertilized ovum is called a ***zygote***

II. Pregnancy

A. General information

1. Pregnancy begins with fertilization and ends with childbirth; this time period, also called *gestation,* averages 38 weeks
   a. The zygote begins to divide as it passes through the fallopian tube; it attaches to the uterine lining by *implantation*
   b. The zygote undergoes a complex sequence of *pre-embryonic, embryonic,* and *fetal development,* finally producing a full-term fetus
2. Because the actual date of fertilization usually is unknown, the expected delivery date typically is calculated from the beginning of the last menstrual period (LMP)
   a. The length of gestation calculated from the LMP is 40 weeks instead of 38 weeks, because the first day of the calculation (the first day of the LMP) is about 2 weeks before ovulation
   b. Gestation length calculated from LMP may be expressed as 280 days (40 weeks × 7 days), as 10 lunar (28-day) months, or as 9 calendar (31-day) months
3. A 9-calendar-month gestation may be subdivided into three periods of 3 months, or *trimesters*

4. Because the uterus grows throughout pregnancy, uterine size provides a rough estimate of the duration of pregnancy

B. Pre-embryonic development

1. The first period of prenatal development, pre-embryonic development begins with ovum fertilization and lasts 3 weeks
2. The zygote undergoes a series of mitotic divisions, or cleavage, as it passes through the fallopian tube
3. The first cell division is completed about 30 hours after fertilization
4. Subsequent divisions occur in rapid succession; the zygote is converted into a ball of cells called a *morula,* which reaches the uterus about 3 days after fertilization
5. Fluid accumulates in the center of the morula, forming a central cavity; the structure now is called a *blastocyst*
6. Blastocyst cells differentiate in two ways
   a. A peripheral rim of cells, called the *trophoblast,* develops; the trophoblast gives rise to the fetal membranes and contributes to placenta formation
   b. A discrete cell cluster in the trophoblast, called the *inner cell mass,* eventually forms the embryo
7. The blastocyst is enclosed in the zona pellucida and remains unattached to the uterus for several days
8. The zona pellucida degenerates during the week after fertilization; this allows the blastocyst to attach to the endometrium and become implanted
9. The inner cell mass becomes a flat structure, called the *germ disc,* which differentiates into three germ layers —the *ectoderm, mesoderm,* and *endoderm;* each layer eventually forms specific tissues in the embryo
10. A cleft appears between the ectoderm of the germ disc and the surrounding trophoblast, forming the amniotic cavity (amniotic sac)
11. A second cavity, called the yolk sac, forms on the opposite side of the germ disc
12. A layer of connective tissue lines the enlarging blastocyst cavity and covers the amniotic and yolk sacs
    a. At this point, the cavity contains the germ disc, amniotic sac, and yolk sac and is called the chorionic cavity (chorionic sac); its wall is called a *chorion*
    b. The entire chorionic sac and developing embryo is called the chorionic vesicle
13. Finger-like columns of cells, called *chorionic villi,* extend from the chorion and anchor the chorionic vesicle to the endometrium

C. Embryonic development

1. During the embryonic period, which lasts from the fourth through the seventh week, the developing zygote begins to assume a human shape and is called an embryo
2. This period of development is critical; all the organ systems form, and the embryo is susceptible to injury by maternal drug use and other factors, such as radiation
3. During this period, each germ layer forms specific tissues and structures in the embryo
   a. The ectoderm primarily forms the external covering of the embryo and the structures that will have contact with the environment

b. The mesoderm forms the circulatory system, muscles, supporting tissues, and most of the urinary and reproductive systems
c. The endoderm forms the internal linings of the embryo, such as the epithelial lining of the pharynx and respiratory and GI tracts

D. Fetal development

1. During the fetal period, which extends from the eighth week until birth, the fetus becomes larger and heavier as it matures; however, it experiences no major changes in its basic structure comparable to those in the embryonic period
2. The fetus displays two unusual features during early development
   a. The head is disproportionately large compared to the rest of the body; this condition changes after birth as the infant grows
   b. The body lacks subcutaneous fat; it fills out shortly before birth, when fat begins to accumulate

III. Structural Development During Pregnancy

A. General information

1. Pregnancy changes the usual development of the *corpus luteum*
2. Various structures develop as a result of pregnancy, such as the *decidua, amniotic sac and fluid, yolk sac,* and *placenta*

B. Corpus luteum

1. For the corpus luteum to function normally, it must be stimulated continually by luteinizing hormone (LH)
   a. Progesterone produced by the corpus luteum inhibits LH release by the pituitary gland
   b. As the corpus luteum ages, it becomes less responsive to LH
   c. Consequently, the mature corpus luteum degenerates unless it is stimulated by progressively increasing amounts of LH
2. When a woman becomes pregnant, placental tissue secretes large amounts of human chorionic gonadotropin (HCG), which is similar to LH and follicle-stimulating hormone (FSH), which also is produced by the pituitary gland
   a. HCG prevents corpus luteum degeneration
   b. It stimulates the corpus luteum to produce large amounts of estrogen and progesterone
3. During the first 3 months of pregnancy, the corpus luteum serves as the main source of estrogen and progesterone; these hormones are needed during pregnancy
4. Later in pregnancy, the placenta produces most of the hormones; although it persists, the corpus luteum no longer is needed to maintain the pregnancy

C. Decidua

1. The decidua is the endometrial lining that has undergone the hormone-induced changes of pregnancy
   a. The *decidua basalis* lies beneath the chorionic vesicle and eventually becomes part of the placenta
   b. The *decidua capsularis* stretches over the chorionic vesicle
   c. The *decidua parietalis* lines the rest of the endometrial cavity

2. The decidua envelopes the embryo and fetus during gestation
3. Decidual cells secrete three substances
   a. The hormone prolactin promotes lactation
   b. The peptide hormone relaxin relaxes the connective tissue of the symphysis pubis and pelvic ligaments and facilitates cervical dilation
   c. Prostaglandin mediates several physiologic functions

D. Amniotic sac and fluid

1. The amniotic sac gradually increases in size; usually by the eighth week of gestation, it completely surrounds the developing embryo and fuses with the chorion
2. The fused amnion and chorion extend from the placental margins to form the fluid-filled amniotic sac, which ruptures at the time of delivery
3. The amniotic sac and its fluid have two important functions
   a. During gestation, they protect the fetus by creating a buoyant temperature-controlled environment
   b. During childbirth, they form a fluid wedge that helps open the cervix
4. Amniotic fluid source and volume vary with the gestational stage
   a. Sources of amniotic fluid are maternal and fetal
      (1) Early in pregnancy, amniotic fluid is derived chiefly from three sources
         (a) Fluid filters into the amniotic sac from maternal blood as it passes through the uterus
         (b) Fluid filters into the sac from fetal blood as it passes through the placenta
         (c) Some fluid also diffuses in the amniotic sac from the fetal skin and respiratory tract
      (2) Later in pregnancy, the fetal kidneys begin to function, adding another source of amniotic fluid
         (a) The fetus urinates into the amniotic fluid
         (b) This urine becomes the major source of amniotic fluid
   b. Maternal and fetal blood filtration and fetal urine excretion continually add to the total amniotic fluid volume
      (1) The volume gradually increases from about 50 ml at 12 weeks' gestation to 800 to 1,000 ml at term
      (2) Amniotic fluid volume reflects the balance between production from maternal and fetal sources and loss through the fetal GI tract
         (a) Normally, the fetus swallows up to several hundred milliliters of amniotic fluid each day
         (b) The fluid is absorbed from the fetal GI tract into the fetal circulation
         (c) Some of the absorbed fluid is transferred from the fetal circulation to the maternal circulation and excreted in maternal urine

E. Yolk sac

1. The yolk sac forms adjacent to the endoderm of the germ disc
2. Part of the yolk sac becomes incorporated in the developing embryo and forms the GI tract
3. Another part of the yolk sac gives rise to primitive germ cells, which migrate to the developing gonads and eventually form oocytes or spermatocytes
4. The yolk sac also forms blood cells during early embryonic development
5. The yolk sac never contains yolk and has no nutritive function; it eventually undergoes atrophy and disintegrates

F. Placenta

1. The placenta is a flattened, disc-shaped structure that weighs about 500 g at delivery; from the third month of pregnancy until childbirth, it supplies nutrients to and removes wastes from the fetus
2. The placenta forms from the chorion (and its chorionic villi) and the adjacent decidua basalis in which the villi are anchored
3. The umbilical cord connects the fetus to the placenta; the cord contains two arteries and one vein
    a. The arteries that carry blood from the fetus to the placenta follow a spiral course on the cord, divide on the placental surface, and send branches to the chorionic villi
    b. Large veins on the placental surface collect blood returning from the villi; these veins join to form the single umbilical vein that enters the cord and returns blood to the fetus
4. The placenta has two circulatory systems
    a. The ***uteroplacental circulation*** delivers oxygenated arterial blood from the maternal circulation to the intervillous spaces (large spaces between the chorionic villi in the placenta)
        (1) Blood spurts into the intervillous spaces from many uterine arteries that penetrate the basal part of the placenta
        (2) Blood leaves the intervillous spaces and flows back into the maternal circulation through veins that penetrate the basal part of the placenta near arteries
    b. The ***fetoplacental circulation*** delivers oxygen-depleted blood from the fetus to the chorionic villi by the two umbilical arteries; it returns oxygenated blood to the fetus by the single umbilical vein
    c. The proximity of the placental circulatory systems brings maternal and fetal blood into close contact; although the maternal and fetal circulations exchange oxygen, nutrients, and wastes, fetal and maternal blood do not mix
5. The placenta produces several peptide and steroid hormones
    a. The two most important peptide hormones are HCG and human placental lactogen (HPL)
        (1) HCG can be detected as early as 9 days after fertilization; the level increases and peaks at about 10 weeks' gestation, then gradually declines
            (a) HCG stimulates the corpus luteum to produce the estrogen and progesterone needed to maintain the pregnancy until the placenta takes over hormone production
            (b) Highly sensitive pregnancy tests can detect HCG in blood and urine even before the first missed menstrual period
        (2) The peptide hormone HPL resembles growth hormone; the level of this hormone rises progressively throughout pregnancy
            (a) HPL stimulates maternal protein and fat metabolism to ensure an adequate supply of amino acids and fatty acids for the mother and fetus
            (b) It antagonizes the action of insulin, decreasing maternal glucose metabolism and making more glucose available to the fetus
            (c) It also stimulates breast growth in preparation for ***lactation***
    b. The placenta produces the two steroid hormones estrogen and progesterone; levels of these hormones rise progressively throughout pregnancy

(1) Estrogen increases uterine muscle irritability and contractility
(a) The placenta produces three different estrogens, which differ chiefly in the number of hydroxyl groups attached to the steroid nucleus; estrone (E_1) has one hydroxyl group, estradiol (E_2) has two, and estriol (E_3) has three
(b) The placenta lacks some of the enzymes needed to complete estrogen synthesis; it requires some precursor compounds produced by the fetal adrenal glands
(c) Because neither the fetus nor the placenta can synthesize estrogens independently, estrogen production reflects the functional activity of the fetus and placenta
(2) Progesterone reduces uterine muscle irritability
(a) The placenta synthesizes this hormone from maternal cholesterol
(b) The fetus plays no part in progesterone synthesis

IV. Labor and the Postpartal Period

A. General information

1. Childbirth ***(parturition),*** delivery of the fetus, is accomplished through ***labor,*** the process in which the fetus is expelled from the uterus by uterine contractions
   a. Weak uterine contractions occur irregularly throughout pregnancy
   b. When labor begins, uterine contractions become strong and regular
   c. Voluntary bearing-down efforts eventually supplement the contractions and lead to expulsion of the fetus and placenta
2. Usually, the head of the fetus occupies the lowest part of the uterus; in this cephalic presentation, the fetus is delivered head-first
   a. Occasionally, the head of the fetus occupies the upper part of the uterus; in this breech presentation, the fetus may be delivered buttocks-, knees-, or feet-first
   b. Other types of presentation are rare, but may include shoulder presentation, in which the fetus is delivered shoulder-first, and compound presentation, in which two presenting parts are delivered first
3. Childbirth is divided into three stages for descriptive purposes
4. The duration of each stage varies, depending on the size of the uterus, the woman's age, and the number of previous pregnancies; for example, the first and second stages commonly are shorter in multiparous women (those who have given birth previously) than in primiparous ones (those who have not given birth previously)
5. After childbirth (postpartal period), the woman returns to the prepregnant state in about 6 weeks

B. Onset and maintenance of labor

1. Several factors contribute to the onset of labor
   a. The number of oxytocin receptors on uterine muscle fibers increases progressively during pregnancy and reaches a peak just before the onset of labor; this increase causes the uterus to become more sensitive to the effects of oxytocin
   b. The uterus stretches as the pregnancy progresses, initiating nerve impulses that stimulate oxytocin secretion from the posterior pituitary lobe

c. The fetus also may play a role in initiating labor
(1) Near term, increased ACTH secretion by the fetal pituitary causes the fetal adrenal glands to secrete more cortisol, which diffuses into the maternal circulation through the placenta
(2) Cortisol increases oxytocin and estrogen secretion and decreases progesterone secretion
(3) These changes in hormone secretion increase uterine muscle irritability and make the uterus more sensitive to oxytocin stimulation
d. Prostaglandins also may play a role in intiating labor
(1) Decreasing progesterone leads to conversion of esterified arachidonic acid into a nonesterified form
(2) The nonesterified arachidonic acid undergoes biosynthesis to form prostaglandins, which stimulate uterine contractions
2. Once labor begins, several factors maintain it
a. Cervical dilation causes nerve impulse transmission to the central nervous system, which increases oxytocin secretion from the pituitary gland
b. Increased oxytocin secretion serves as a positive feedback mechanism; it stimulates more uterine contractions, which further dilate the cervix and cause the pituitary to secrete more oxytocin
c. Oxytocin also may stimulate prostaglandin formation by the decidua; prostaglandins diffuse into the uterine myometrium and enhance contractions

C. Stages of labor
1. The first stage of labor is characterized by cervical effacement (thinning) and dilation; the fetus begins to descend
a. Before labor begins, the cervix is not dilated; by the end of the first stage, the cervix is dilated fully
b. Uterine muscles contract actively while the cervix and the lower part of the uterus thin and dilate
c. The amniotic sac and fluid function as a hydrostatic wedge to help dilate the cervix
d. The first stage lasts from 6 to 24 hours in primiparous women; it is much shorter in multiparous women
2. The second stage covers the time between full cervical dilation and expulsion of the fetus
a. Uterine contractions increase in frequency and intensity, and the amniotic sac ruptures
b. As the flexed head of the fetus enters the pelvis, pelvic muscles force the head to rotate anteriorly and force the back of the head (occiput) under the symphysis pubis
c. Uterine contractions force the flexed head deeper into the pelvis
d. The resistance of the pelvic floor gradually forces the head into extension
(1) As the head presses against the pelvic floor, the vulvar tissues stretch, and the anus dilates
(2) At this stage, the vulvovaginal orifice usually is enlarged surgically by a small incision, called an episiotomy
e. As the head is delivered, the face passes over the perineum, and maternal tissues retract under the chin
f. The head, which had been rotated anteriorly during fetal descent, rotates back to its former position after passing through the vulvovaginal orifice

g. Usually, the head undergoes lateral (external) rotation as the anterior shoulder rotates forward to pass under the pubic arch
h. Delivery of the shoulders and the rest of the fetus follows shortly afterwards
i. This stage averages about 45 minutes in primiparous women; it may be much shorter in multiparous women

3. The third stage of labor begins immediately after childbirth and ends with expulsion of the placenta
    a. After delivery of the neonate, the uterus continues to contract intermittently and reduces in size
    b. The area of placental attachment also is reduced correspondingly; because the bulky placenta cannot decrease in size, it separates from the uterus
    c. Blood seeps into the area of placental separation, the decidua basalis
    d. As the uterus continues to contract, retroplacental blood is compressed and acts as a fluid wedge, cleaving the placenta from the uterus
    e. Usually, the edges of the placenta are the last to separate from the uterine wall; the midportion of the placenta, covered by the fetal membrane, commonly is expelled first
    f. Less commonly, the lower edge of the placenta separates and is expelled first
    g. This stage averages about 10 minutes in primiparous and multiparous women

D. Postpartal period

1. After childbirth, the reproductive tract requires about 6 weeks to return to its former condition
    a. The uterus rapidly decreases in size; most of the involution occurs in the first 2 weeks after delivery
    b. The stretched tissues of the pelvis and vulva return to their former state more slowly
2. Postpartal vaginal discharge *(lochia)* persists for several weeks after childbirth
    a. *Lochia rubra* (bloody discharge) occurs from 1 to 4 days postpartum
    b. *Lochia serosa* (pink-brown, serous discharge) occurs from 5 to 7 days postpartum
    c. *Lochia alba* (white, brown, or colorless discharge) occurs from 1 to 3 weeks postpartum

V. Lactation

A. General information

1. *Lactation,* milk production by the breasts, is regulated by the interactions of four hormones
    a. Estrogen and progesterone are produced by the ovaries and placenta during pregnancy
    b. Prolactin and oxytocin are produced by the pituitary under hypothalamic control; some prolactin also is produced by decidual cells during pregnancy
2. Placental production of estrogen and progesterone increases progressively throughout the pregnancy, stimulating proliferation of glandular and ductal tissue in the breasts
3. Prolactin causes milk secretion after the breasts have been stimulated by estrogen and progesterone

4. Prolactin secretion increases throughout pregnancy in response to high estrogen levels; paradoxically, the high estrogen and progesterone levels that occur during pregnancy inhibit milk secretion by the prolactin-stimulated breast
5. Oxytocin from the posterior pituitary lobe causes contraction of specialized cells, which help expel milk during breast-feeding
6. Breast-feeding stimulates prolactin secretion; the resulting high prolactin level causes changes in the mother's menstrual cycle

B. Hormonal initiation of lactation

1. Progesterone and estrogen levels fall sharply after childbirth when the placenta is expelled
2. Because estrogen and progesterone no longer inhibit the effects of prolactin on milk production, the mammary glands begin to secrete milk
3. Prolactin secretion also decreases after delivery unless the nipples are stimulated by breast-feeding
   a. Sensory impulses from the nipples are transmitted to the hypothalamus; this stimulates prolactin release from the anterior pituitary lobe
   b. Milk secretion continues as long as the nipples are stimulated regularly by breast-feeding; if breast-feeding is discontinued, the stimulus for prolactin release is removed and milk production stops
4. Breast-feeding also stimulates oxytocin
   a. Sensory impulses from the nipples are transmitted to the hypothalamus; this causes oxytocin release from the posterior pituitary lobe
   b. Oxytocin causes contraction of the myoepithelial cells surrounding the breast lobules
   c. This contraction causes *milk ejection* (expulsion of breast milk from the secretory lobules into larger ducts)
   d. The breast-feeding infant readily obtains milk from these ducts

C. Effects of breast-feeding on the menstrual cycle

1. The high prolactin level in a postpartal woman inhibits FSH and LH release
2. If a woman does not breast-feed her infant, prolactin output soon declines and FSH and LH production by the pituitary no longer is inhibited
   a. Cyclic release of FSH and LH soon follows
   b. Normal menstrual cycles commonly resume about 6 weeks after delivery or a few weeks after discontinuation of breast-feeding
3. If the woman breast-feeds her infant, the menstrual cycle does not resume because prolactin inhibits the cyclic release of FSH and LH necessary for ovulation; consequently, a breast-feeding woman usually does not become pregnant
4. The amount of prolactin released in response to breast-feeding gradually decreases
   a. The inhibitory effect of prolactin on FSH and LH release also declines
   b. Consequently, ovulation and the menstrual cycle may resume; pregnancy may occur after this, even though the woman continues to breast-feed her infant

Study Activities

1. Discuss the importance of spermatozoa transport to ovum fertilization.
2. Trace the growth of the fertilized ovum through the pre-embryonic, embryonic, and fetal development stages.
3. Describe the effects of pregnancy on the corpus luteum.
4. Identify the substances secreted by the decidua and decribe their functions.
5. Describe the functions of the amniotic sac and fluid.
6. Compare the uteroplacental and fetoplacental circulations of the placenta.
7. Identify the placental hormones and their functions.
8. Describe the role of the fetus in inducing labor.
9. Discuss the positive feedback mechanism that maintains labor.
10. Explain the physiologic events associated with each stage of labor.
11. Describe how hormones initiate lactation.

Appendix

Selected References

Index

Appendix: Glossary

Accommodation —adjustment of the lens of the eye to change the focal length and bring images into sharp focus on the retina
Acid-base balance —stable concentration of hydrogen ions in body fluids
Active transport —movement of a substance across a cell membrane by a chemical activity that allows the cell to admit larger molecules than would otherwise be able to enter; this transport mechanism requires energy
Aerobic metabolism —energy-releasing process that requires oxygen, in which glucose is converted to pyruvate and then oxidized to yield carbon dioxide, water, and adenosine triphosphate (ATP) by mitochondrial enzymes
Agglutination —clumping
Air conduction —sound wave transmission through the tympanic membrane and auditory ossicles
All-or-none response —response that governs muscle fiber contractions in a motor unit; a nerve impulse strong enough to stimulate contraction causes all the fibers to contract
Anabolism —synthesis of simple substances into complex ones
Anaerobic metabolism —energy-releasing process that does not require oxygen, in which glycogen and glucose are broken down into lactic acid
Binocular vision —vision that produces a three-dimensional image, which is necessary for normal depth perception
Bone conduction —sound wave transmission through the bones of the skull
Buffer system —system that minimizes pH changes caused by excess acids or bases (alkalies)
Capacitation —activation process for spermatozoa that allows ovum penetration; small perforations appear in the acrosome (head cap) of the spermatozoon, allowing enzyme release
Cardiac cycle —events that occur during a single systole and diastole of the atria and ventricles
Cardiac output —amount of blood ejected per minute from a ventricle, which equals the stroke volume (volume of blood ejected from a ventricle at each contraction) multiplied by the heart rate in beats per minute
Catabolism —breakdown of complex substances into simpler ones or into energy
Chemotaxis —movement toward or away from a chemical stimulus
Citric acid cycle —metabolic pathway by which a molecule of acetyl-CoA is oxidized enzymatically to yield energy
Coagulation —blood clotting
Complement system —collection of about 20 proteins in blood and tissue fluids that, when activated, augments the effects of antibodies
Concentration gradient —differences in the concentration of the molecules on each side of a cell membrane
Condensation reaction —chemical reaction in which molecules join to form a more complex compound; the reaction commonly causes the release of a molecule of a simple compound, such as water (H_2O)
Conditioned reflex —reflex that can be learned as a result of past associations; for example, salivation induced by the sight or smell of specific foods
Convergence —eye alignment so that the image of a near object falls at corresponding points of both retinas, allowing depth perception
Countercurrent mechanism —process by which the kidneys concentrate urine
Cross bridges —structures formed by the binding of myosin filament heads to actin filaments; they pull actin filaments toward the center of the sarcomere, causing muscle contraction
Cross over —mixing of genetic material in meiosis; during synapsis, chromatid

segments of homologous chromosomes break off and interchange
Deamination —removal of the amino group -NH_2 from a compound
Depolarization —change in resting membrane potential toward zero
Diastole —cardiac relaxation
Diffusion —movement of dissolved particles (solute) across the cell membrane from one solution to a less concentrated one
Digestion —breakdown of food into absorbable nutrients in the GI tract
Ejaculation —pulsatile expulsion of semen from the penis; it is associated with the erotic sensations called orgasm
Ejection fraction —amount of blood ejected during each ventricular contraction in relation to the end-diastolic volume
Electrolyte —substance that dissociates into ions when dissolved in water and can conduct an electrical current
Electron transport system —metabolic pathway that converts products of the citric acid cycle into energy
Emission —semen movement into the urethra
End-diastolic volume —total blood volume in each ventricle before ventricular systole
Endochondral bone formation —type of ossification in which a mass of cartilage forms, is invaded by osteoblasts, and is converted into bone
End-systolic volume —unejected blood that remains in each ventricle
Equilibrium —sense of balance
Erythropoiesis —red blood cell production
Expiratory reserve volume —amount of air that can be exhaled forcefully at the end of a normal tidal expiration (about 1,300 ml in the average adult)
Extracellular fluids compartment —intravascular fluid (in blood plasma and lymph) and interstitial fluid (in loose tissue surrounding the cells)
Facilitated diffusion —type of diffusion involving a carrier molecule in the cell membrane that picks up the diffusing substance on one side of the membrane and deposits it on the other side
Fertilization —union of a spermatozoon and ovum
Fetoplacental circulation —circulatory system that delivers oxygen-depleted blood from the fetus to the placenta by the umbilical arteries; it returns oxygenated blood to the fetus by the umbilical vein
Fibrinolysis —blood clot dissolution
Fight-or-flight response —response to stimulation of the sympathetic nervous system; it prepares an individual to cope with stress; also called alarm response
Focal length —distance from a convex lens to its focal point
Focal point —point behind a convex lens at which parallel light rays are brought into focus
Functional areas —parts of the cerebral cortex related to specialized functions
Gas exchange —process of respiration in which gases are transported between the air and capillaries in the lungs
Gluconeogenesis —glucose synthesis from amino acids
Glycogenesis —glycogen synthesis from glucose
Glycogenolysis —glycogen breakdown into glucose
Glycolysis —metabolic pathway that converts glucose to pyruvic acid or lactic acid and yields energy
Hematopoiesis —formation and development of blood cells from stem cells
Hemostasis —arrest of bleeding by complex mechanisms that include vasoconstriction and coagulation
Heterozygous —having different alleles at a given locus on homologous chromosomes
Homozygous —having identical alleles at a given locus on homologous chromosomes
Hydrolysis —chemical reaction in which a chemical bond is broken and the atoms of a water molecule are added across the break. Hydrogen is added to one side, and the hydroxyl group to the other side

Hydrostatic pressure —pressure that filters fluid from the blood through the capillary endothelium
Immunity —ability to resist organisms or toxins that can damage tissues
Inspiratory reserve volume —amount of air that can be inhaled forcefully at the end of a normal tidal inspiration (about 3,000 ml in the average adult)
Interstitial fluid —fluid surrounding the cells
Intracellular fluid compartment —fluid in the body's cells
Intramembranous bone formation —type of ossification in which osteoblasts form bone without a preliminary cartilage mass
Intrapleural pressure —pressure in the pleural cavity (space between the lung and chest wall)
Intrapulmonary pressure —air pressure in the lungs
Labor —process in which the fetus is expelled from the uterus by uterine contractions; it is divided into three stages
Lactation —milk production by the breasts, which is regulated by estrogen, progesterone, prolactin, and oxytocin
Lipogenesis —lipid synthesis from glucose
Meiosis —special type of cell division in gametes, in which the daughter cells receive half the number of chromosomes of the parent cells
Menarche —onset of menses
Menopause —cessation of menses, resulting from ovarian follicle depresion; typically occurs around age 45
Menstrual cycle —recurring cycle in which a layer of the endometrium is shed, and then regrows, proliferates, and sheds again
Metabolism —transformation of substances into energy or materials the body can use or store; consists of the two processes anabolism and catabolism
Milk ejection —expulsion of breast milk from the secretory lobules into larger ducts of the breast
Mitosis —type of cell division in which the parent cell produces identical daughter cells with chromosomes duplicated from the parent; it occurs in all human cells, except gametes
Negative feedback mechanism —mechanism in which an elevated level of a hormone or hormone-regulated substance suppresses further hormone output
Neuromuscular junction —motor endplate and axon terminal associated with it
Oogenesis —process of ova formation
Opsonization —process by which phagocytosis increases
Optic tracts —bundles of optic nerve fibers that connect the optic chiasma and the brain stem
Orgasm —climax of sexual excitement usually accompanied by seminal fluid ejaculation in the male
Osmolarity —measure of osmotic pressure exerted by a solution
Osmosis —movement of water molecules across the membrane from a dilute solution (having a high concentration of water molecules) to a concentrated one (having a lower concentration of water molecules)
Osmotic pressure —pressure exerted by a solution on a semipermeable membrane
Ovulation —ovum release from an ovarian follicle
Oxidation —loss of electrons by a chemical compound
Pain —sensation of discomfort or suffering caused by pain receptor stimulation
Partial pressure —pressure exerted by a single gas in a mixture of gases; it is designated by the letter "P" preceding the chemical symbol for the gas
Parturition —delivery of a fetus; childbirth
Passive transport —movement of small molecules across the cell membrane by diffusion; this transport mechanism requires no energy
Peripheral resistance —degree of impedance to blood flow, which may be increased by vasoconstriction

Peristalsis —movement of the contents of a tubular organ through wavelike, rhythmic muscular contractions
pH —measurement of the hydrogen ion concentration in body fluids; a pH of 7 is neutral, less than 7 is acid, and more than 7 is alkaline
Phagocytosis —process by which specialized cells engulf and dispose of organisms, other cells, and foreign particles
Polarization —state of a nerve fiber not transmitting an impulse; the fiber exterior has a positive charge, the interior a negative charge
Puberty —developmental stage in which reproductive ability begins and secondary sex characteristics develop
Reduction —gain of electrons by a chemical compound
Referred pain —internal organ pain that is perceived as coming from the body surface at a site distant from the organ
Reflex —automatic or involuntary action in response to a stimulus, usually mediated through the spinal cord
Reflex arc —chain of sensory, connecting, and motor neurons that produces a reflex when stimulated
Refraction —bending of light rays as they pass from one medium to another of different density
Refractive index —measurement of the refractive power of a medium; the greater the refractive index, the more the light rays are bent as they pass through the medium
Refractory period —brief period after nerve impulse transmission during which the nerve fiber is unresponsive and cannot transmit another impulse
Renin-angiotensin-aldosterone mechanism —self-regulating mechanism that helps control blood pressure, blood volume, and sodium plasma concentration
Residual volume —amount of air left after the lungs expel the expiratory reserve volume (about 1,200 ml in the average adult)
Resorption —bone breakdown, which normally equals bone formation during adulthood
Respiratory center —control center in the brain stem that regulates the rate and depth of respiration
Resting membrane potential —voltage difference between the interior and exterior of a nerve fiber
Reuptake —absorption of a neurotransmitter into the synaptic vesicles of the axon terminals that originally released it
Saltatory conduction —type of conduction in myelinated nerves in which the depolarization wave jumps between gaps in the myelin sheath rather than moving progressively along the fiber, as in unmyelinated nerves
Sexual response —physiologic effects of sexual stimulation, which include penile erection and semen discharge in the male, and erection of the clitoris and labia minora and uterine and tubal contractions in the female
Sodium-potassium pump —active transport mechanism that carries potassium ions in through the cell membrane and simultaneously transports sodium ions out
Spermatogenesis —process of spermatozoa formation
Starling's law —the amount of stretching of cardiac muscle fiber helps regulate stroke volume
Stroke volume —volume of blood ejected from a single ventricle with each contraction
Synapse —small gap between two neurons or between a neuron and an effector organ
Systole —cardiac contraction
Tidal volume —amount of air inhaled in one breath (about 500 ml in the average adult)
Transamination —exchange of an amino group in an amino acid for a keto group in a keto acid, through the action of transaminase enzymes
Tropic hormone —substance that stimulates hormone production by a target gland

Uteroplacental circulation —circulatory system that delivers oxygenated arterial blood from the maternal circulation to the placenta by the uterine arteries; it returns blood to maternal circulation by veins near the arteries
Ventilation —process of respiration in which air moves in and out of the lungs
Vital capacity —maximum amount of air that can be moved in and out of the lungs by maximum inspiration and maximum expiration (about 4,800 ml in the average adult)
Voiding reflex —involuntary reflex that causes the urge to urinate when the bladder contains 300 to 400 ml of urine; it is under a high degree of voluntary control
Zygote —fertilized ovum

Selected References

Anthony, C.P., and Kolthoff, N.J. *Textbook of Anatomy and Physiology,* 13th ed. St. Louis: C.V. Mosby Co., 1989.

Ganong, W.F. *Review of Medical Physiology,* 15th ed. Los Altos, Calif: Lange Publishing Co., 1991.

Guyton, A.C. *Textbook of Medical Physiology,* 7th ed. Philadelphia: W.B. Saunders Co., 1986.

Seeley, R.R., et al. *Anatomy and Physiology,* 2nd ed. St. Louis: Times-Mirror/Mosby College Publishing, 1989.

Tortora, G.J., and Anagnostakos, N.P. *Principles of Anatomy and Physiology,* 6th ed. New York: Harper & Row, 1990.

Vander, A.J., et al. *Human Physiology: The Mechanisms of Body Function,* 5th ed. New York: McGraw-Hill Book Co., 1990.

Index

i refers to an illustration; t, to a table

D

E

i refers to an illustration; t, to a table

i refers to an illustration; t, to a table

i refers to an illustration; t, to a table

i refers to an illustration; t, to a table

Notes

Notes

Notes